UINTAH RAILWAY
Pictorial

BALDWIN LOCOMOTIVE WORKS PHOTO – H. L. BROADBELT COLLECTION

UINTAH RAILWAY No. 50 was ready to be shipped from the Baldwin Locomotive Works of Philadelphia. This specially designed narrow-gauge articulated engine was the "offspring" of Uintah management.

UINTAH RAILWAY
Pictorial

VOLUME I

Mack to Atchee

BY RODGER POLLEY

SUNDANCE
Books

UINTAH RAILWAY *Pictorial*

— VOLUME I —
Mack to Atchee

BY RODGER POLLEY

For My Dad...
Donald Polley

In Memory of Mom...
Bonnie Laurel Polley

SUNDANCE PUBLICATIONS *Ltd.*
250 BROADWAY • DENVER, COLORADO 80203
Telephone 303 / 777-2880

Published by...
Sundance Publications, Limited
Denver, Colorado

Production Manager — Dell A. McCoy
Editor and Typographer — Russ Collman
Photographic Technician — Steven Meyers
Scanner Technician — Theo Berlyn
Editorial Consultant — Kenton Forrest

ISBN 0-913582-68-9

AL AND BERNADETTE CHIONE COLLECTION

FRONT COVER —
Original Art By Howard Fogg...

This beautiful painting of the Uintah Railway's articulated locomotive No. 51 was produced by the famous railroad artist Howard Fogg. This scene is based on the photograph taken by Otto C. Perry, reproduced on Pages 152–153.

BACK COVER —
The Division Point of Atchee...

Atchee, Colorado, was situated where the graded automobile road disappeared in the distant valley. The Uintah Railway gained elevation to climb over Baxter Pass, where the photographer stood to take this view. This scene represents the steepest railway grade for an adhesion line in Colorado.

DELL A. McCOY PHOTO

CHRIS H. UEBERROTH COLLECTION

Acknowledgments

THE LIST OF PEOPLE who have helped in so many ways to make this book possible could go on for pages. Help has been freely given from so many individuals that it is impossible to remember all of the small contributors that have made this a reality over the past eight years.

First and foremost has been the support of my family — for without their encouragement, I would not have been motivated enough to keep this project going. As a topic of conversation, they have all listened to as many hours of Uintah Railway history as any non-railfan would want to endure. Vernon Powell, Christy Lancaster, Becky and Marvin Noel, as well as Robert and Mary Ann Flynn — thank you. Also, thanks go to my father, Donald Polley, for his support on every level. As a child, it was his interest in the old ghost towns around the mines at Dragon that peaked my interest in the past.

No list could be complete without mentioning the Uintah's foremost historian, Henry Bender. His original work, *Uintah Railway – The Gilsonite Route,* was the original source of interest in this relatively small narrow-gauge railroad. The enlightening information and the well laid out format of the book have been essential to anyone's attempt at adding more chronological data to an already well-thought-out and complete history.

Nevertheless, like any chronicle of events, there always are more pieces of the puzzle out their waiting to be found. The most interesting of those pieces have been the people who actually worked and lived in the many small towns along the Uintah route. Nearly 60 years have passed since the Uintah Railway's trains ran through thriving little communities that are now long gone, and at this writing, the number of individuals who are still with us is gradually becoming smaller. With their insight, memories and photographs you can almost grasp the way things existed when steam-driven trains were still the most widely used means of transportation in use.

Grateful thanks go to Elizabeth Angus, Harry Babcock, Helen Banks, Kelly Barlow, John C. Beaslin, Peter Bennett, Janis Bigalo, Hazel Boardman, Gordon Booth, Stella Boren, Betty Cosseboom, Donald and Virginia (Bathurst) Compton, Roy Eno Jr., Mildred and Jim Finnicum, Paul and Betty Garber, Dell Jacobsen, Marie Kaczmerek, George Komatas, Frank Mancuso, Kay Martin, James McCoy, Homer Sonney Phillips, Porter Phillips, Sparky Rector, Florence Reynolds, Margret Romager, Otis Simpson, Earl "Dick" Wagaman, Carlyle Workman, Ernie Workman, Arthur Hillman, George Roseland and William O. Larson.

Before his death, Richard A. Ronzio had planned to produce a photographic book featuring his extensive collection of Uintah Railway photographs. His family has aided the publisher with additional historical pictures of the Uintah Railway before his collection was sold. Special mention also should be made regarding the late Robert C. Boulter, whose photographs and writings are contained in this volume.

The ***Uintah Railway Pictorial*** would never have come to pass if it was not for the careful preservation of the many old photographs from the Uintah Railway's time period. Thanks go to all the people who were nice enough to give their historic pictures for use in this book. Gratitude goes to Olive Blackburn, John and Hazel Carson, Richard and Francis Dewey, Lawrence Deved, Cliff Haycock, Arthur Hillman, William C. Larson, Bill Lee, John W. Maxwell Jr., Derrin Noel, Charles Pollock, Jack Pollock, Dan Price, Francis Rizzari, George Roseland and Chris Ueberroth.

Thanks also go to the many professional people and their staffs, who helped with the acquisition of images and printed information from the various public collections around the United States. My first contact with a public collection of photographs came with my introduction to Doris Burton in the Uintah County Library of Vernal, Utah. In an age of "voice mail," "faxes," "web sites," detailed request forms, Doris Burton's hands-on friendly approach, coupled with her genuine interest in the subject, helped so much in those early years of research. And it is certain she will help again as the research continues on the north end of the Uintah Railway.

Gratitute goes to Raymond Clark of the Denver Public Library, the staff of the Colorado Historical Society, Ellen Halteman and Sarah Logue from the California State Railroad Museum, Yavonne Peterson of the Lower Valley Historical Society (in Fruita, Colorado), and archivist David Pfeiffer of the National Archives and Records Administration provided assistance in obtaining copies of the informative Interstate Commerce Commission (ICC) Report, as well as copies of ICC maps. A special thanks goes to Judy Prossor Armstrong, from the Museum of Western Colorado, who freely gave so much time and effort to this project — being "blessed with the affliction" of being the nearest museum (just down the street), with a sizable collection of photographs, maps and other information the author could constantly refer back to.

Thanks also to Bill Ellicot, Harry Babcock and Warren Keifer for technical assistance — as well as for their work and the work of other members of the Rio Grande Chapter of the National Railway Historical Society (NRHS) at the Cross Orchards Living History Farm in Grand Junction, Colorado. Their work and the work of those at the Colorado Railroad Museum in Golden, Colorado, have kept some of the Uintah's history alive by keeping some of its old equipment in tip-top condition.

Living only a few miles from where the Uintah Railway once operated, has allowed the author the opportunity to travel over the old grade many times in search of the location of certain photographs, or the whereabouts of the sites of long-gone structures or bridges. Many friends and relatives have accompanied the author on these trips, helping to make the days more interesting, with their insights regarding how things might have looked 60-plus years ago. Thanks go to John Carson for help in tracking down the bridges and ghost towns on the south side of Baxter Pass.

Also worthy of mention are H. A. Warren and Cliff Haycock, who provided two antique railroad pieces, which allowed the author to trade with collectors for more photographs that had not been seen before. Much of the section of the book about the railroad yards at Atchee would not have been included if it were not for the selfless acts of these men.

Special credit goes to my brother-in-law, Terry Lancaster. His never-ending interest in the Uintah Railway and the ghost towns it left behind has been an inspiration to gather even more information.

And last, but not least, is my wife, Debbie. Through our entire relationship, there has been never-ending support — in so many ways. Whether it was the many trips to the desert or Baxter Pass, the complete taking-over of our conversations, or hours spent on the word processor, thank you with all my heart.

—— Rodger Polley, Grand Junction, 1999

Table of Contents

Uintah Map List

Introduction

Most people who have been associated with railroads, especially historians, railfans and folk from the Rocky Mountain region, have heard the Uintah Railway name. Famous because of two of the largest narrow-gauge locomotives ever built, as well as for the legendary 7.5-percent grade up Baxter Pass, the unique name of this railway is still recognized 60 years after its demise in 1939. This publication has been designed to provide as detailed a look into the past as possible, via as many photographs as could be found. This garnering of images, both from my personal collection, as well as from many generous people who have loaned their pictures for use in the book, has grown to such an extent that one volume could not contain everything that has been discovered about this railway and the small towns it created.

Volume I of the ***Uintah Railway Pictorial*** follows the right-of-way of the railway, as it stretched out to the north of Mack, Colorado. First crossing a nearly barren desert, and then entering the rugged canyons of the Bookcliff Range, the Uintah stopped at the small coal-mining settlement of Carbonera and then continued on up the canyon to Atchee, Colorado. At Atchee, most of the railway's activity happened on a daily basis, as this small town was used as a main division point and engine-servicing facility for the entire line. It is the history in photographs of this section of the line that this volume hopefully will bring to light.

Pronounced YEW-IN-TAH, this word has come from the Ute Indian culture of northeastern Utah. "Uintah" and "Uinta" both are pronounced the same way, and they have been used to name all sorts of things in this part of the country. A mountain range, a river, a tribe of the Ute nation and a county government — along with countless other political and geographic locations — have taken this name.

ROBERT C. BOULTER COLLECTION

SAM ATCHEE was one of the chiefs of the Uintah Ute tribe. His name was chosen for the Uintah Railway's division point: ATCHEE. Even though the U.S. Congress had agreed upon an area in Utah designated as the Uintah Reservation, Bert Seabolt, assistant to the chief engineer and manager of construction for the Denver & Rio Grande Railway between 1882 and 1890, became the negotiator with Congress and principal promoter of Gilsonite mining on reservation land. His efforts in 1888 allowed part of the Uintah Reservation to be taken away from the Utes—the part with the Gilsonite ore. The Act of Congress permitting this transgression provided the Uintah Ute Indians with a mere $20.00 per acre—truly a steal, considering the true worth of the land and the limited knowledge the Indians had of what was going to take place. It should be remembered that all of Utah and western Colorado originally and rightfully belonged to the Ute nation.

The Uintah Railway came into existence as a corporation on November 4, 1903, and used the name to give the railway a local flavor, aiding in the fascination of the public at the time in the American Indian culture. All the promoters of the railroad were capitalists by any definition, caught up in the industrial revolution that was sweeping the United States at the turn of the century. As all of the big railroads in the U.S. were simply named after the cities and locales that they were built to and from, i.e., Colorado & Southern, Denver & Rio Grande, Chicago & Northwestern, etc., the Uintah Railway was named like so many other small railways, with a certain flare to attract attention and the all-important passenger business. The whole reason for one of the last narrow-gauge railways to be built in the United States was the unusual natural resource found only along the northern portion of the border between Colorado and Utah.

Gilsonite, as it would be known, was a naturally occurring asphaltite substance found in mineable quantities only in one place in North America, and perhaps in the whole world. At first glance, someone might easily mistake a pile of this substance as coal, being black and somewhat fractured in the same way. However, on closer inspection, the fracturing of the Gilsonite usually is smaller and thinner, and not as "chunky" as sedimentary coal. The surface of a piece of this material generally is glossy and smooth, and when scratched against something, leaves a brown streak rather than a black mark, as associated with coal or other asphaltic substances.

More important than its physical appearance, though, are the unique chemical properties of the Gilsonite. A low melting point and light weight, along with superb insolative qualities, make it a highly valuable commodity still in demand to this day. As different and interesting as Gilsonite is to the chemist and industrialist, so, too, was the unusual nature of the way it occurred to the geologist and miner. Large horizontal layers of sedimentary sandstones and shales had through the ages of geologic time fractured vertically. These fractures were then filled-in with some sort of hydrocarbon liquid, which over time, gradually lost the more volatile and lighter liquids, leaving behind the nearly pure carbon structure of Gilsonite. The fractures often were several hundred feet deep and could run for miles in a southeast-by-northwest direction. This formed a continuous vein of the material that could be from only a few inches wide to over 15 feet in width in some veins north of the White River. The tops of these large veins of Gilsonite were eroded away in most places, exposing the

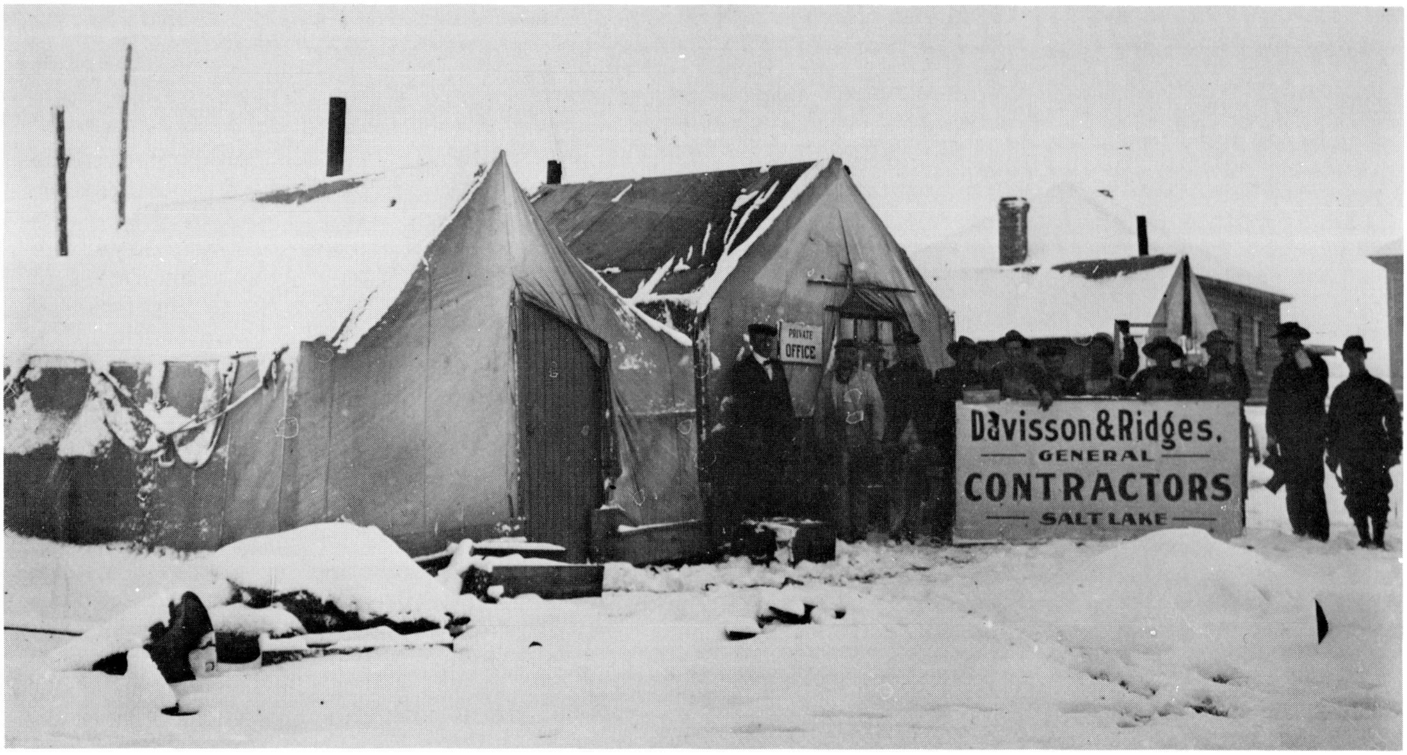

FRANK A. KENNEDY PHOTO – RICHARD A. RONZIO COLLECTION

A LARGE GROUP of laborers and engineers stood in the snow in Mack, Colorado, during the early months of 1904. The Utah Construction Company was hired as the main contractor for grading and building the Uintah Railway. Davisson & Ridges was one of the many subcontractors that built homes, offices, hotels and installed the many lines for water, sewer and steam service. This concerted effort allowed little towns like Mack to appear virtually overnight. From surveys made the preceding year, the Uintah Railway's mainline, company towns and coal mine were all constructed and completed within 10 month's time.

entire length of the vein as it crossed over hill and valley into the distance. Although today, artificial means can be used to get the various substances that can be obtained from Gilsonite, they are not as economically feasible as mining it from the earth. At the turn of the century, there were no substitutes for the many and varied uses of Gilsonite, and mining it was to become a very profitable venture.

Gilsonite got its name from one of its earliest promoters and entrepreneurs, Sam Gilson. Scientifically, "Gilsonite" was named "uintahite" as early as 1885, but the Gilsonite name would end up sticking with it as it was promoted by Sam Gilson and his business partners, especially after the first mining company was named the Gilsonite Manufacturing Company in 1888. This original company was bought out by a group of investors from Missouri, and the new company was named the Gilson Asphaltum Company in 1889. This company was again bought out in 1900, and it retained the Gilson name. By 1903, the General Asphaltum Company owned all of the stock in the Gilson Asphaltum Company, and combined with the opening of more of the Uintah Indian Reservation to mining, the time was ripe for the mining of Gilsonite to commerce on a scale never seen before.

Early mining in the late 1880's had provided evidence that through a simple procedure, a varnish could be obtained from the Gilsonite that would become highly valuable to industry. Continued research by Sam Gilson and his partner, Bert Seaboldt, found more uses and markets for the "new" hydrocarbon with every ton of the ore that was removed from the Carbon Vein mines west of the Green River, near Fort Dueshane, Utah. Political lobbying was done in Washington, D.C., for the Carbon vein to be taken out of the Uintah Indian Reservation, and after that was accomplished, most of the Gilsonite that was produced came from these mines. Across the border in Colorado, a few shallow veins of the valuable ore were found and exploited, but for the most part, most of the mineable claims were still located inside the reservation.

Due to their location inside the Uintah Indian Reservation, the mining claims there had very little assessments or improvements made to mineable areas. Most smaller companies and individuals who did have claims there, would not see any money from their efforts as long as the lands stayed inside the reservation. Charles O. Baxter was president of the Gilson Asphaltum Company through the 1890's, and during that time, he had slowly accumulated many of the claims for the Gilsonite veins in the district. With more political pressure, Congress finally opened the land to mining activities in 1903. Mr. Baxter quickly bought up most of the remainder of the claims from individuals, leaving the Gilson Asphaltum Company with more than 90 percent of the mineable veins in northeastern Utah.

The biggest problem and largest expense involved in getting Gilsonite to market was the need for a wagon haul from the Carbon Vein to Price, Utah — or from the veins in western Colorado to the Denver & Rio Grande Railroad at Rifle, Colorado. Not only was the expense too high, but the amount of ore was too small, and the trip took too long to economically transport Gilsonite to as many markets as possible. The obvious solution was to construct the most efficient form of transportation available for moving large amounts of ore — a railroad. The Barber Asphalt Paving Company was the parent company of the Gilson Asphaltum Company, and it would be their capital that would be used to move the Gilsonite to market. Negotiations with the Denver & Rio Grande Railroad to build a feeder line from the mines in Utah were unsuccessful. The Barber company then was faced with the reality that if a railroad was to be built, they would have to build it.

The nearest mineable deposits of Gilsonite to the mainline of the Denver & Rio Grande were at the southern-most vein of the Dragon Mine. As a result, mining operations were immediately switched to concentrate on the Dragon Mine. At the same time, information about the territory in which the feeder line was to build had to be discerned. Surveying commenced immediately in the area to figure out the logistics for building a railroad. Even an uneducated man of the time would have quickly figured out that the terrain was difficult for constructing a railroad. Early hopes that the railroad could be built to standard gauge to eliminate the need for transferring ore from one railroad gauge to another were quickly dashed by the engineering reports that were received. The tracks would have to be laid down as narrow gauge — three feet between the rails, rather than the standard 4 feet, $8^1/2$ inches that was hoped for. Standard-gauge trackage could have been attained, but the construction cost would have been at least doubled, as earth work and bridges would have been increased, as would the length of the proposed railroad — extended to attain the same elevation as a narrow-gauge line would require.

Only 53 miles from the D&RG line, the Gilsonite ore at the Dragon Mine was not all that far away; however, what a trip it was going to be to transport it that distance. The Bookcliff Mountains ran in an east–west direction, perpendicular to the proposed location of the railroad. Unlike the mountain ranges most people think of when they consider the mountains of Colorado — huge granite snowcapped peaks, with well-eroded valleys below the crags — the Bookcliff Mountains were altogether different. Geologic pressures from inside the Uintah Basin to the north had caused a huge swell of earth to push upward over time, forming a continuous chain of mountains, from near Rifle, Colorado, to Helper, Utah. Huge chunks and layers of sandstone and shale had been broken and pushed upward and over one another, forming a jagged topography of cliffs, ledges and boulders.

It was through this terrain that the builders of the Uintah Railway were going to have to construct the line to move the ore to market. Following the natural grade of the drainage streams on both sides of the mountain range, a preliminary location for the railroad was decided upon. From the originating point on the Denver & Rio Grande Railroad, near a siding known as Crevasse, Colorado, at an elevation of 4,541 feet, the new railroad was laid out, following the basic course of West Salt Wash Creek for 28 miles. At that point, the course of the wash was left behind in order to get to the lowest point at the top of the mountain pass.

This portion of the line would prove to be one of the most sensational pieces of railroading to be built in the United States — with a 7.5-percent grade ruling the line for the next five miles, as the railroad wound its way up the side of what would soon be known as Baxter Pass. Here, at an elevation of 8,437 feet, the grade of the railroad descended along the north side of the same mountain at a steady 5.0-percent grade for the next 12 miles. There, it followed the decline of Evacuation Creek, as it flowed north toward the White River. Fifty-three miles later, and at an elevation of 5,788 feet, the railroad terminated at the mining town of Dragon, Utah.

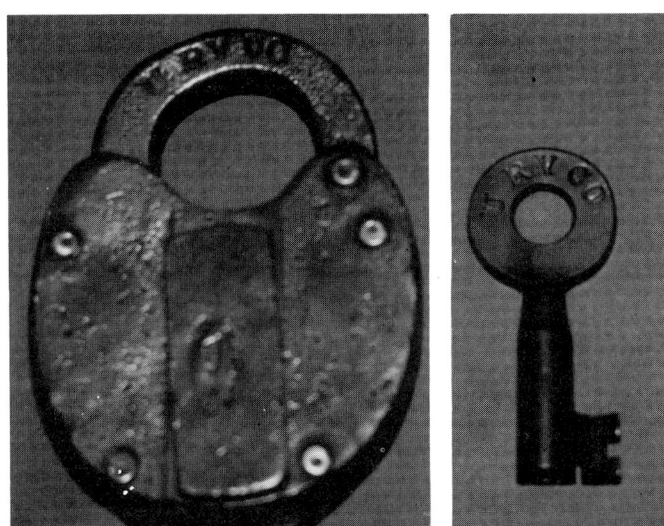

CHRIS H. UEBERROTH COLLECTION

SWITCH LOCK and KEY

— THE UINTAH RAILWAY —

And Connections

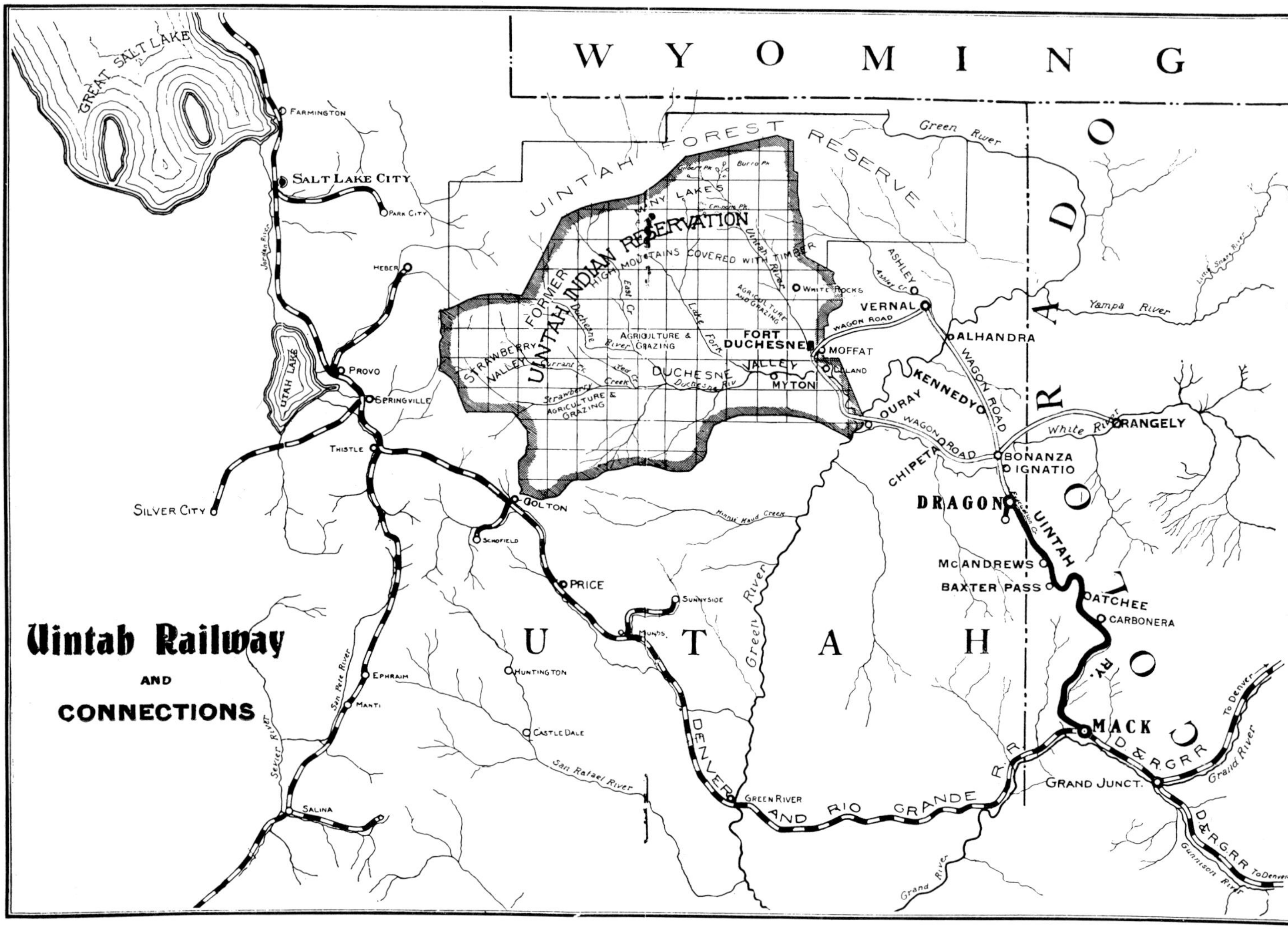

FRANCIS AND FREDA RIZZARI COLLECTION – COURTESY ED AND NANCY BATHKE

A PRE-1911 PAMPHLET produced for advertising, illustrated the orientation of the Uintah Railway on a map showing parts of Colorado, Utah and Wyoming. Labeled "Uintah Railway and Connections," it indeed shows this narrow-gauge short-line railway and all of the connections that the Uintah had with other railroads, namely the Denver & Rio Grande Railway at Mack, Colorado (in the lower right-hand corner), and the wagon road terminating at Dragon, Utah. This dirt road fanned out into the Uintah Basin to serve Ouray, Fort Duchesne and Vernal, Utah, as well as Rangely, Colorado, with stops at various stagecoach stations along the way.

As with most railroad ventures during the period before and just after the beginning of the 20th century, there was always an eye for future expansion of the Uintah Railway, as seen here with the many "agriculture and timber" references in the area. At this time, the Uintah line had not yet built its extension to yet-to-be-settled mining town of Rainbow, Utah, just north of Dragon. Even though it would never build a line across the White River into northeast Utah (in the area between Bonanza and Vernal), the Uintah Toll Road made the railway the natural choice for passengers and freight coming and going to the entire district.

From Beginning to End

WITH THE GRADES and curves needed on Baxter Pass, a narrow-gauge railroad was required to haul Gilsonite ore from the mines in Utah to the railroad connection in Colorado. Mining had been going on during all the summer of 1903 at the mine at Dragon, Utah, and the ore was transported over Baxter Pass by horse and wagon to the Denver & Rio Grande Railway, at the siding called Crevasse, Colorado. The railway surveying was completed that summer and fall, and the Uintah Railway came into existence on November 4, 1903. Contracts for building the line were awarded by the end of the month, and construction would continue through that first winter and end nearly a year later, when the first train pulled into Dragon, Utah, during October of 1904.

Unlike most of the other mining railroads that were built in the Rocky Mountains to tap some natural resource, the Uintah Railway was unique in that unlike other railroads built to serve the many private mining operations of an area, the Uintah's parent companies owned nearly all of the mines that had production possibilities. Instead of bustling boom towns full of private ventures from mines and tourist destinations, the territory in which the Uintah Railway was operating was one of the last huge tracts of land in the West that was all but deserted. Isolated, remote, arid and unimproved, the terrain in which the Gilson Asphaltum Company's mines were located was inhospitable to human work. Not only would the Uintah Railway be building a difficult-to-operate narrow-gauge railway, but the line would have to provide every other piece of civilization to the miners and railroaders who were filling the ranks of the company's payroll, as well as the traveling public.

Building a completely isolated railway, especially to a non-standard gauge, meant there would be no chance to lease equipment, or otherwise rely on a connecting or neighboring railroad for help. Aside from the right-of-way and structures that made the railway run, there were many necessities required for life in this area. Foremost was the need for water, not only for the steam locomotives that relied on it, but also for the people living in this arid region. Great expense was taken to provide water tanks at several locations on the railway, and running pipe lines to feed them from natural springs on Baxter Pass, as well as at Wendella and Atchee. This need for water would also be evident in operating expenses over the years, as countless water cars full of the precious liquid would be transported from one place to another along the Uintah's right-of-way.

Housing for the employees was another matter that would have to be dealt with. Once again, there was no other private venture like a hotel or boarding house at any of the locations along the railway where an employee could lodge. The Uintah Railway would have to build houses for its employees, not just out of the human necessity, but out of the simple fact that most railroaders would not take a position with the railway if there was not going to be a hospitable place to stay. The traveling public also had to be thought of, as plans for the new railway were put down on paper. Some kind of housing for non-employees would have to be provided for the many travelers that would step off the trains and have no place to stay. However, the Uintah would not shrink from its responsibilities.

The Uintah Railway was not going to be a railway that was built by scrimping on pennies. It was the favorite child of a group of wealthy investors and companies back East, and it would continue in that role until near the end. For the superstructure of the railway, there was always an eye toward sound business decisions; however, at the same time, everything that was needed, was acquired. Special Shay geared engines were bought to do the demanding job on Baxter Pass, while unusual 0-6-2T tank engines were used exclusively for passenger trains. A complete machine shop would be built at the new company town of Atchee on the south side of the pass, with every piece of modern equipment needed to rebuild locomotives and rolling stock.

The employees were taken care of, also, as new cottages for workers were built at all of the new towns, along with matching buildings for all of the section houses. Special residences for the superintendent and general manager, ice houses (with free ice), free coal and company-subsidized school buildings created a livable, but far from lush environment, to live and work in. The public also was looked after, especially since they were paying, with three hotels along the

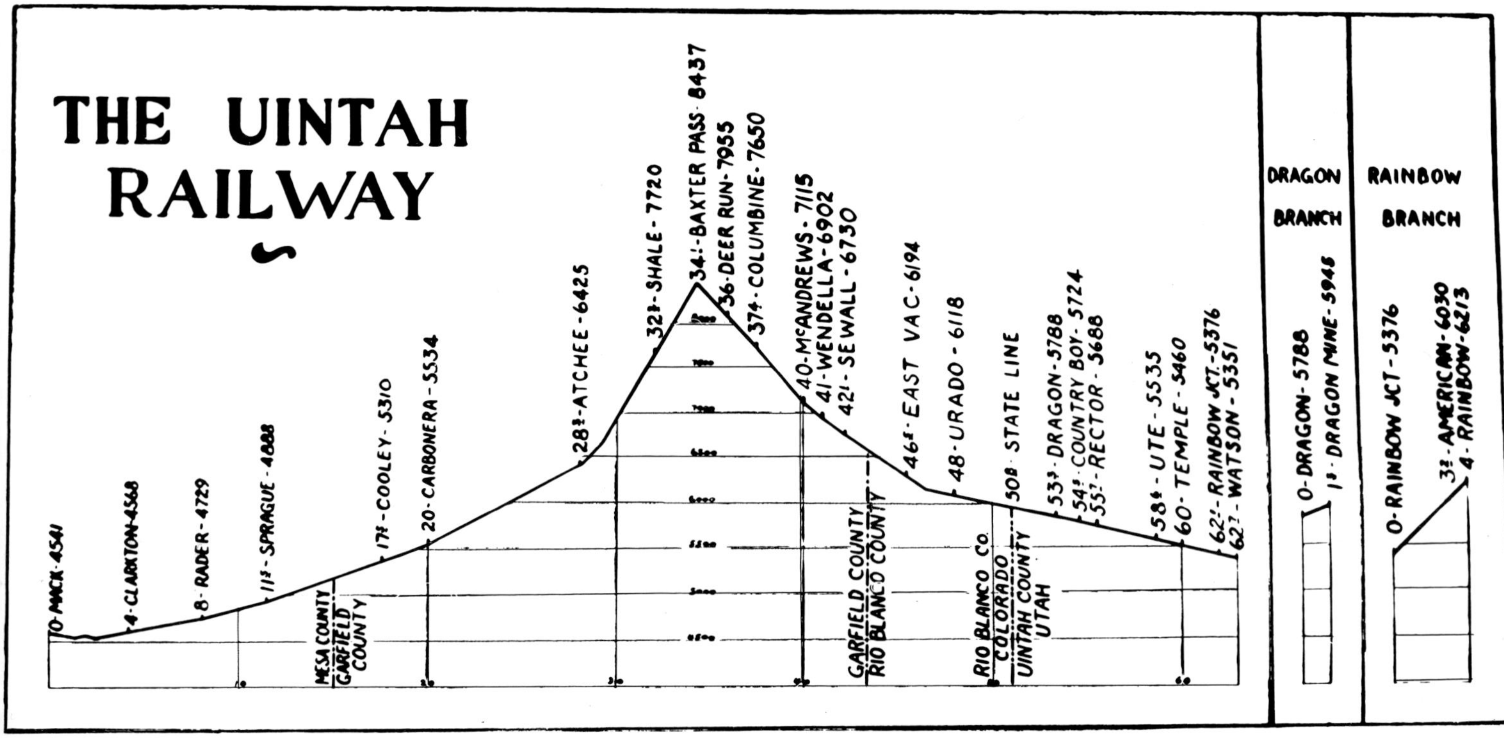

CHRIS H. UEBERROTH COLLECTION

line, as well as the creation of a toll-road system and freight company at the north end of the railway, to the towns of northeastern Utah.

So... the Uintah Railway was off with a bang! Money from the parent corporation went into building this shortline railway to their mines in northeastern Utah, getting the ball rolling for an ever-growing industry.

The Uintah Railway went through four distinct periods in its history. The first one was the time period of its infancy, from 1904 through 1910. As it always would be, Gilsonite ore was the largest single commodity shipped on the railway. After all, this was why the Uintah was built. During these first years, almost all of the production of Gilsonite was coming from the Cumberland Tunnel at the Black Dragon Mine. Geologically different from all the other veins of Gilsonite in the area, the Black Dragon Mine had a lot of suspended sandstone and other debris in the ore, which caused it to be mined differently at times than the rest of the Gilsonite mines. It was here in the early years that the Uintah Railway experimented with transporting the black hydrocarbon that had been mined in loose form. Although it took fewer man hours to pack it in bags and transfer it to standard-gauge cars, it soon was found impractical for at least two reasons. First and foremost was the fact the Gilsonite could have different chemical properties, depending on where it was found in the vein. This, in turn, created a quality scale, which buyers used to determine its chemical use and therefore its value. Bagging it at the mine by its content was the only sure way of defining its chemical properties and value. Of a more practical matter for the railway was the increased risk of fire, associated with the transportation of the highly flammable ore to Mack, the Uintah terminal on the Denver & Rio Grande mainline. These two factors resulted in the total cessation of loose-ore transportation by 1910.

From its inception, the railway was basically operated by using three divisions. These divisions were: [1] the run from Mack to Atchee, [2] the climb over Baxter Pass, and [3] the trackage going to the mines around Dragon, and later, Rainbow. During the first period of time, the Uintah Railway's fleet of locomotives reached no more than seven: two small tank engines for passenger traffic, two 2-8-0 Consolidations for the freight traffic on the south end of the line, and three Shay geared engines, used to climb the stiff grade and traverse the sharp curves on Baxter Pass. This was a simple operation, as the single mine at Dragon produced most of the ore, and operations in and out of the district were not complicated. Gilsonite from the mine and travelers to and from the Vernal, Utah, area, along with the freight shipments into the small farm communities of the Uintah Basin, via the stagecoach road, was all the traffic the Uintah Railway had to deal with.

By the end of the first decade, both through passenger and freight traffic was increasing with the passing of every month. A large fire in the Cumberland Tunnel of the Black Dragon Mine had caused a potential problem with operations and railway revenues in February of 1908, but by the end of that year, production from the new Country Boy Mine, west of the Black Dragon, and limited mining at the Cumberland and Black Diamond claims led to further increases in the total traffic on the Uintah Railway that year.

Over the next two years, mining was again going on at the Black Dragon Mine, and by 1910, mining operations had left few mineable deposits near the railway at the Dragon Mine. That year, some production was accomplished out of the Temple Mine farther along the vein system to the west, and like the Country Boy Mine, the wagon haul to the railhead was beginning to be a problem. The continuously increasing demand for the hydrocarbon ore left no other choice than for the Uintah to expand north toward more deposits of Gilsonite ore.

The year 1911 brought in the next period in the Uintah Railway's history, which would last for the next 10 years. Expansion to the mines at Rainbow, Utah, was the next logical step, as the mines near Dragon were beginning to play out. The new track was laid down Evacuation Creek to the new "railhead" at Watson, Utah, where the wagon road went on to Vernal. Five miles to the west, the Uintah Railway's tracks crawled up a steady 5.0-percent grade to the level of the Gilsonite veins located at Rainbow. Production from mining in the Rainbow area would dramatically increase the railway's revenues over the next several years, and these were the halcyon days for the Uintah Railway.

The fleet of locomotives almost doubled during this period, increasing to 13 units by February of 1920, trying to keep up with the ever-increasing flow of ore from the mines and the equally impressive passenger and freight business in and out of the Uintah Basin towns. Freight cars also were increased during this period, creating even more work for the shops at Atchee, as more men were put to work to deal with the ever-increasing traffic. It was during this period that the talk of the Uintah's extension across the White River to the north, to the large deposits of Gilsonite on that side of the river, while the possibility of an eventual extension into Vernal, Utah, was heard the most.

Those plans came close to fruition in 1921, as surveys were completed on the extension to the Bonanza veins north of the White River. Unfortunately, that year ended up to be the turning point for the destiny of the railway, as a national recession took its toll on the demand for Gilsonite. The previous decade had been looked upon with great expectations and grand designs, but with operating revenues deep in the red in 1921, the plans to expand north of the White River were all but forgotten.

The decade of the Roaring Twenties would turn out to be another defining period for the railway. During that 10-year span, the management of the company changed, and an emphasis on thrift and efficiency was placed on the railway that had never existed in its growing years. Financial trouble from operating debts that occurred during the recession of 1921 caused the Uintah Railway's management to continually try to increase efficiency and do away with the old flamboyancy of the early years. Coupled with the fact that automobiles and motor trucks would become a widely used form of local transportation in the coming years, the 1920's ushered in a pragmatism through the Uintah's management, as well as the management people of the corporations that controlled it, that although the Uintah was still the most effective way to transport Gilsonite, it would have to earn that right against ever-increasing forces of competition and the inevitable depletion of ore in the mines on the south side of the White River. However, for students of railroading, the ever-increasing push for efficiency led to a most remarkable project. In 1926, and again in 1928, responding to the needs of a narrow-gauge carrier that was "making it work," while most other narrow-gauge railroads in the Rockies were fading into obsolescence, the Baldwin Locomotive Works created two remarkable pieces of machinery to ascend the grades on Baxter Pass. Unlike anything else before or after them, these narrow-gauge articulated locomotives were designed to go around 66-degree curves on a 7.5-percent grade, while pulling nearly as much freight as two of the Shay engines they replaced. These two large 2-6-6-2T locomotives breathed new life into the "bottom line" of the railway and were the epitome of the many projects that kept the machine shops busy in Atchee and the books in the black into the early 1930's.

The final stage of the Uintah Railway's life was spent in the terrible Great Depression years — just getting by. No matter how well the railroaders did their job, there was one fact that they could do nothing about. The Gilsonite mines on the south side of the White River were quickly running out of mineable deposits of ore. Trucking the Gilsonite from remote mines at Bonanza, Utah, on the north side of the river, was becoming a necessity, and by 1938, the economies associated with trucking the Gilsonite to the standard-gauge railroad (the Denver & Salt Lake Railway) at Craig, Colorado, were much less than the expensive proposition of first transferring it from the trucks to Uintah narrow-gauge cars and then transferring it again to standard-gauge cars for shipment to market. Coupled with this situation was the fact that the Uintah Railway was a wholly owned subsidiary and was totally dependent on the traffic of ore, and the railway now was a losing proposition.

The Uintah Railway was abandoned in 1939, and all of its equipment and support system was dismantled and torn down — leaving only the scars of the abandoned right-of-way. In a few months, the little railway was gone, taking less time for removal than it took to appear on the scene nearly 35 years earlier. Beyond that, it took the livelihoods of many people with it, including communities that had once prospered because of the railway operation.

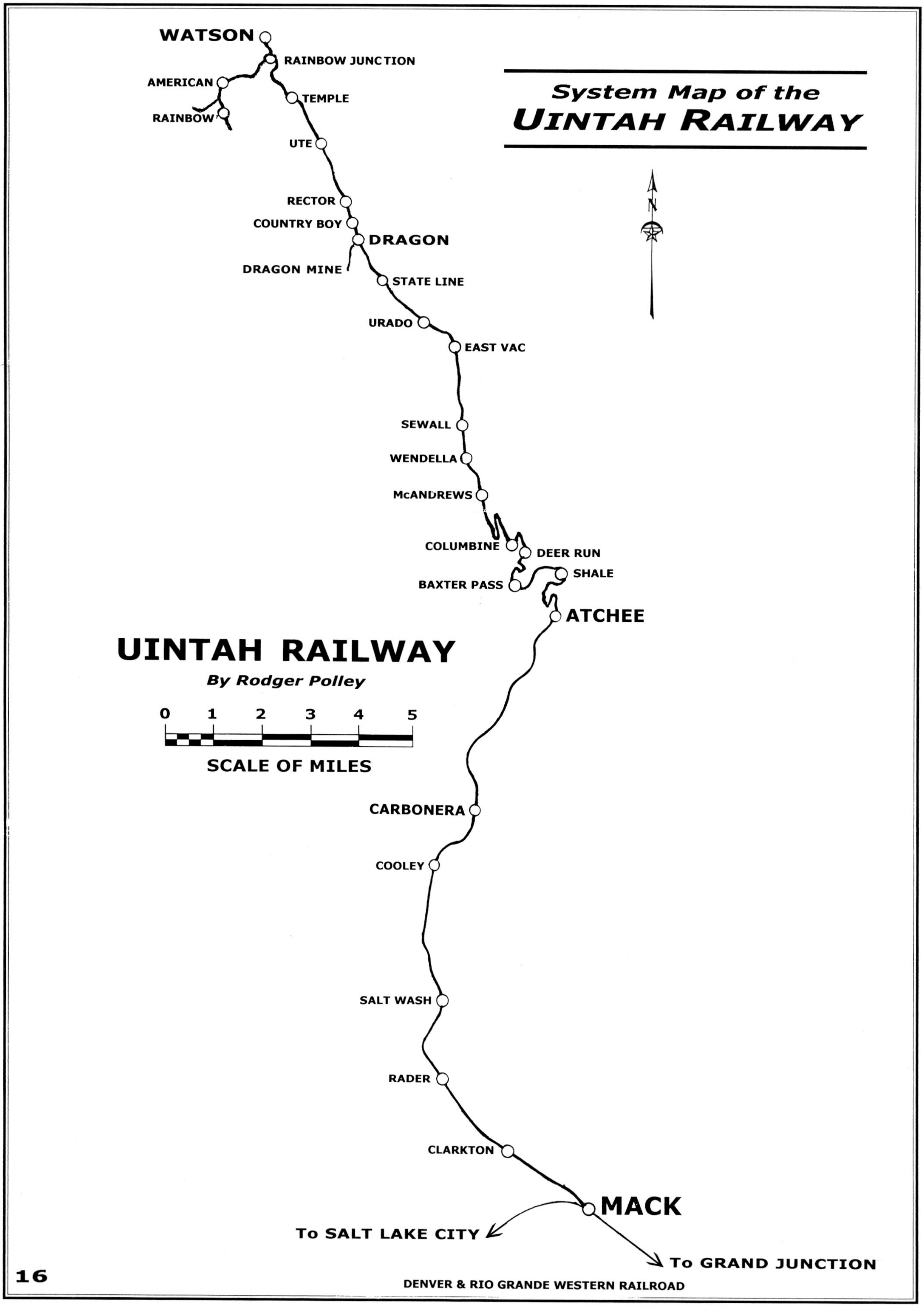
System Map of the
UINTAH RAILWAY
WATSON
RAINBOW JUNCTION
AMERICAN
RAINBOW
TEMPLE
UTE
RECTOR
COUNTRY BOY
DRAGON
DRAGON MINE
STATE LINE
URADO
EAST VAC
SEWALL
WENDELLA
McANDREWS
COLUMBINE
DEER RUN
SHALE
BAXTER PASS
ATCHEE
UINTAH RAILWAY
By Rodger Polley
0 1 2 3 4 5
SCALE OF MILES
CARBONERA
COOLEY
SALT WASH
RADER
CLARKTON
MACK
To SALT LAKE CITY
To GRAND JUNCTION
DENVER & RIO GRANDE WESTERN RAILROAD

Profile of Uintah Railway

— *Mack to Atchee* —

March of 1930

By Rodger Polley

THIS PROFILE MAP shows the grade of the Uintah Railway trackage from Mack to Atchee, Colorado. To the far right is the vertical scale by elevation (in feet) above sea level, while the horizontal scale is below (in miles). The grade is given in percentage (%) on top of the line, while the rise in feet over the same sections is given in feet below the line. The names of stations and sidings along the line are given with their mile marker first, followed by the elevation. Bridge locations and their corresponding numbers are given in the next section down. The Alignment Guide gives the location and degrees of curvature on the line. Notice how tight and frequent the curves became just out of Atchee. The next line down shows the alignment and location of sidings, how many cars each siding could hold, wye and an open O for water and a black ● for coal. The last line shows rail weights at various points along the railway line.

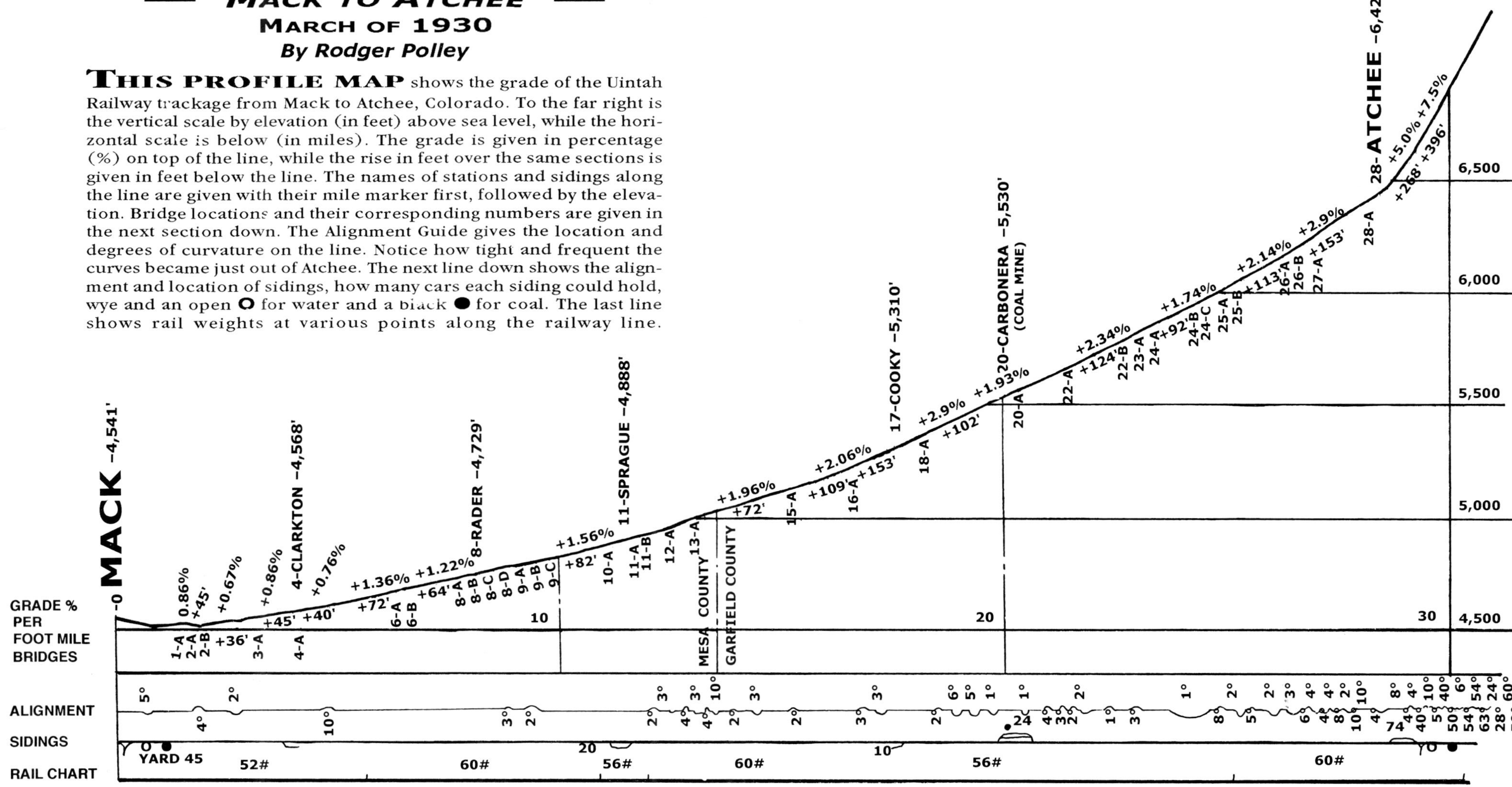

W. C. LARSON PHOTO COLLECTION

UINTAH ENGINE No. 11 (the second No. 11) was pulling away from the depot in this early photograph of Mack, Colorado. She was on her way toward Atchee, with a mixed freight-and-passenger train. Like all of the other Consolidation engines, this 2-8-0 was unable to negotiate the 7.5-percent grade on Baxter Pass with more than three cars. This engine was equipped with blind center drivers that could make it around the sharp 66-degree curve at Morro Castle; however, she lacked adequate tractive effort, so No. 11's longest run was on the 28 miles from Mack to Atchee. Notice the stark landscape, as it appeared during the early days.

THIS GROUP OF PEOPLE seemed to be in high spirits as they "waited for the train" on the platform at the Mack depot in 1905 or 1906. The man in the center of the photograph had lightened-up the usual somber attitude that surrounded having one's picture taken by putting on a woman's cap for the cameraman. The picture was taken from the Denver & Rio Grande's mainline, while the Uintah's office building is in the background. (See railway map on Page 52.)

FRANK A. KENNEDY PHOTO – RODGER POLLEY POSTCARD COLLECTION

Chapter 1

The Rio Grande Connection... MACK, COLORADO

DURING THE SUMMER of 1903, in the desert 16 miles west of Grand Junction, Colorado, a small group of surveyors was busy gathering supplies for a big job. Over the next few months, this group of men would survey for a narrow-gauge railway that would extend to the north for 52 miles from this location on the Denver & Rio Grande Railroad. It would go over the rugged Bookcliff Mountains to the Gilsonite mines in northeastern Utah. Before their arrival, this location was a sparsely inhabited region, with only a few hearty farmers and ranchers living in the immediate vicinity. During the late 1890's, three brothers, Frank, Ben and Joe Kiefer pooled resources to extend an irrigation ditch from the west end of the Grand Valley Canal to the area around what would become Mack, Colorado. A few farmers ventured into this desert area and utilized the small towns of Loma and Fruita, Colorado, as their sources for supplies, as well as markets for their produce.

The survey crew had chosen this spot because it was the most northwest section of the Denver & Rio Grande Railroad system in this area of Colorado. The Gilsonite mines were northwest of here, over the Bookcliff Mountains, and this was the nearest location to the mines. In 1890, the Denver & Rio Grande had converted its mainline from narrow-gauge to standard-gauge trackage, and during the change-over had moved their line south from the desert to the bank of the Colorado River. At Mack, the new standard-gauge D&RG route broke away from the desert and curved sharply into a small ravine carved out by Salt Wash Creek, down to the level of the Colorado River, where it continued for several miles, until it came back out of the river canyon near Cisco, Utah. The location at the future site of Mack was a natural place for the small town that would provide facilities to transfer the many thousands of tons of Gilsonite ore and other freight to and from the narrow-gauge railway being surveyed. The survey crew quickly drew up plans for the small railroad town in the desert — and marched off on the old abandoned D&RG narrow-gauge roadbed, not knowing the many engineering problems that lay ahead. William Banks, a future railway station agent and foreman for the Uintah Railway at Watson, Utah, was a laborer on the survey crew, and years later, he recalled the sense of adventure and accomplishment that accompanied the men on their journey over the Bookcliff range.

By the end of October 1903, the survey crew had completed their work at the northern terminating point of the new railway's right-of-way, at the Black Dragon Gilsonite mine in Utah. On November 4, 1903, the Uintah Railway was incorporated in Colorado, and by November 14, the new company had awarded a contract to the Utah Construction Company of Salt Lake City, Utah, to build the narrow-gauge railway. Within a few weeks, the Denver & Rio Grande Railroad was receiving shipping orders to deliver freight to Crevasse, Colorado. Crevasse was the name of the station and passing siding on the D&RG adjacent to the Mack location. The siding at Crevasse was improved, and the Utah Construction Company began to move men and materials to Mack. As the winter progressed, detailed surveys of the first few miles of grade were completed, and other crews were hard at work surveying the sites where the many wooden trestles were needed to cross West Salt Wash Creek.

The railway terminal and settlement of Mack were named for John M. Mack, who was the president of the General Asphalt Company and its subsidiary, the Barber Asphalt Company, and now, the new Uintah Railway Company. The town of Mack was quickly rising out of the desert, and businessmen in Grand Junction and Fruita were stunned to find out over the next few months that another railway was being built in Mesa County, without much fanfare. The parent General Asphalt Company had no interest in drawing attention to an already politically charged situation regarding the legal issues around the Gilsonite claims on the Uintah Indian Reservation. The reservation was opened to the public in 1903, and at this time, there was considerable controversy concerning the ownership of Gilsonite claims that had been filed prior to the opening to the public. With their almost-total monopoly of the Gilsonite claims in Utah, the General Asphalt Company wanted to complete the railway and begin meaningful ore production with as little controversy as possible.

The hustle and bustle of a railway construction camp could not be ignored. The word got out through the Grand Valley that another rail line was being constructed, and individuals made their way to Mack to look over the prospects of creating business in the area. Carl Osborne was a grocer in Fruita, Colorado, and he decided to build the Fruita Mercantile Company store at the east end of Mack in 1904. This name would later be changed several times, but the store

OGDEN 309 M — MACK ELEV 4540 — DENVER 469 M

was commonly known as the "Mack Mercantile." Other than this store, there were few other people who could spare the capital to invest in the budding community.

Except for this small privately owned store, Mack basically was going to become a company town. As the spring of 1904 progressed, and the grading crews and tracklayers made their way northwest into the desert, the general contractors awarded several contracts for the building of a number of structures in Mack. The large hotel, company offices, water tank, boilerhouse and employee houses were paid for by the Uintah Railway. It is interesting to note the great expenses the Uintah company absorbed, in comparison with most other mining railways in Colorado. All of the buildings erected during the original construction period had a kind of grand scale to them. Most of Colorado's mountain railroads were built from a busy town to an already-developed mining community. The Uintah Railway literally was building a railway from nowhere to nowhere.

Until May of 1904, the little town of Mack was home to hundreds of horses and mules that churned up the dust as they hauled the many bridge timbers, railroad ties, steel rail and other supplies needed by the work crews that were strung out all along the Uintah Railway's right-of-way. New loads of supplies were being unloaded by the Denver & Rio Grande Railroad at the Mack yard almost daily. By the end of the month, the Uintah Railway's first engine had been purchased secondhand and shipped to Mack. This Baldwin Consolidation was purchased by the Uintah to serve on work trains while two new engines were being constructed. This first engine was a former D&RG 2-8-0, No. 55, named the *Tomichi*. She remained as No. 55 for awhile, but later was renumbered as No. 11. As Mack really started to look like a railway town, construction on the rest of the line now proceeded quickly, as supplies were moving faster to the railhead.

The transfer tracks in the Mack yard would never again see the large volume of materials and equipment being transferred from the D&RG to the Uintah, or vice versa. Every tie, timber, spike and piece of hardware imaginable was being unloaded at Mack and transported north to the railhead and the little company towns along the way. Over the next two years, the Uintah Railway received 12 gondolas from the National Dump Car Company. Also received were freight cars from the American Car & Foundry Company in St. Louis, Missouri, including 12 flatcars, 8 boxcars and 5 stock cars — a total of 37 cars. Although built for the sole purpose to haul Gilsonite ore, the Uintah Railway was designated as a common carrier. By midsummer, the Lima Locomotive Works' Shay No. 1, and the Baldwin Locomotive Works' Consolidation No. 10 had been delivered to Mack, and they were put to service helping build the railway. After being fired-up and sent north on the newly constructed trackage, Shay No. 1 would not be seen very often on the south end of the railway. For a few months, this geared locomotive was the only engine on the roster that could maneuver on the steep grade and sharp curves on Baxter Pass.

By October of 1904, the Uintah Railway had reached Dragon City, Utah, 53 miles northwest of Mack. The Mack terminal was still extremely busy, as new employees were hired every day at the company offices and were given assignments at different locations all along the railway. New equipment continued to be unloaded at the transfer tracks, including Shay No. 2, fresh from the erecting floor in Lima, Ohio. On

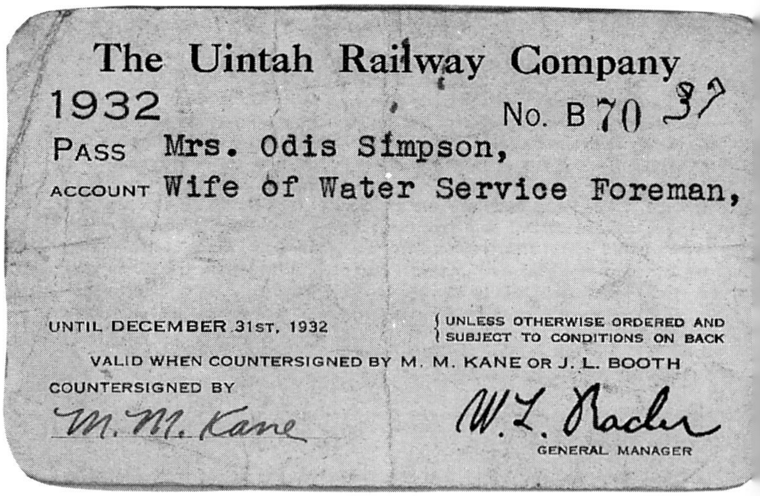
The Uintah Railway Company
1932 No. B 70 37
PASS Mrs. Odis Simpson,
ACCOUNT Wife of Water Service Foreman,
UNTIL DECEMBER 31ST, 1932 — UNLESS OTHERWISE ORDERED AND SUBJECT TO CONDITIONS ON BACK
VALID WHEN COUNTERSIGNED BY M. M. KANE OR J. L. BOOTH
COUNTERSIGNED BY M. M. Kane
W. L. Rader
GENERAL MANAGER

RODGER POLLEY COLLECTION

top of all of this incoming freight, the laborers at the transfer tracks got their first taste of the heavy black bags of Gilsonite ore, which would make up the majority of the freight hauled by the Uintah Railway for the next 34 years. The Black Dragon Mine had been stockpiling Gilsonite ore for several months, to save freight costs of the wagon-haul to the railway. Immediately, the stockpile of Gilsonite ore was made priority No. 1, and the Uintah's four engines and their new crews met the challenge.

It is interesting to note the speed at which Mack, Colorado, and the Uintah Railway were engineered and built. Within a year, Mack went from nothing but a sagebrush flat to a fully staffed and functional railway town. A wye for turning the engines around, a boilerhouse, a water tank that supplied each house with running water, a sewer system, a railway hotel, a railway office building, a joint D&RG/URY depot, a hardware and grocery store, an enginehouse, employee houses and an icehouse were designed and built in the time it takes a modern bureaucracy to do a "site plan and study."

From 1904 until 1910, the people of the little town of Mack concentrated on running a smooth-operating railway. The year 1905 saw the delivery of more freight cars and three new engines. Shay No. 3 was purchased for use on Baxter Pass and the north end of the line for freight trains, while two sister engines, Nos. 20 and 21, were small 0-6-2T tank engines built by Baldwin Locomotive Works to haul the Uintah's passenger trains over Baxter pass to Dragon City, Utah. Three former D&RG Pullman sleeping cars also were purchased in 1905, to provide accommodations for the under-estimated passenger traffic.

Even though the Uintah Railway had only penetrated the Uintah Basin as far as Dragon City, there was an ever-increasing use of the Uintah's passenger service to the towns in the basin. The opening of the land in the former Uintah Indian Reservation brought many prospectors and farmers to northeast Utah in search of riches and productive land. Most of the settlers were quickly disenchanted by what they actually found, and they either never staked a claim or moved elsewhere after only a short stay.

While many of the people who bought tickets at the Mack depot were curious adventurers out to see what was available on the old reservation land, the great majority of the Uintah's passenger traffic came from Vernal, Utah. Vernal was the largest town in Uintah County and was the center for trade for most of the smaller agricultural towns in the region. Although the Uintah Railway stopped short of Vernal by more than 50 miles, it was now the fastest and most efficient form of transportation in and out of the Uintah Basin for both passengers and freight.

In their 1905 advertising booklet called ***THE UINTAH RAILWAY,*** the Uintah's management described Mack as follows:

> Mack is in the broad, fertile valley of the Grand River, 20 miles westward from Grand Junction, Colorado, and, while a small community, is yet destined to grow to considerable importance as a railway town, surrounded, as it is, by land unexcelled in fertility when brought under irrigation.
>
> The Uintah Railway Company, for the convenience of its patrons, has constructed most comfortable hotels both at Mack and Dragon. Harmonizing with the country in which they are placed, they are in what might be termed the *Adobesque Renaissance* style of architecture, with walls of concrete. The lounging-rooms and parlors are equipped with books, current magazines and daily papers, and the dining-rooms most inviting and the bedrooms models of cleanliness and comfort. Lighted by gas, heated by steam, with the necessary luxuries of tub and shower baths and with most excellent cuisine, there is abundant invitation to both traveler and tourist to not only come, but linger. The railway, hotels and connecting lines are operated throughout the year.

By the end of 1905, the main construction of most of the buildings in Mack was completed. The majority of the men who came to work for the Uintah Railway late in 1904 and 1905 did not bring their families with them, and they did not arrive until the men were settled-in on their new jobs and had acquired houses. Over the course of these two years, enough families had moved to Mack to justify a school by 1906. The Uintah offered a small yard-office building to be used as the school for $1.00 per year rent. This first school building was east of the Uintah's wooden water tank, in the middle of the transfer yard. During the following year, the school was moved to its permanent location next to the Mack Mercantile store.

For the next few years, Mack's railway traffic and the growth of the town remained about the same. By 1910, things were changing, along the Uintah Railway and at the Gilsonite mines at Dragon, Utah. During 1908, fires at the Black Dragon Mine and the continued depletion of the ore veins, forced the Uintah Railway and its parent company, the Barber Asphalt Paving Company, to begin looking for more distant sources of Gilsonite ore. By 1910, continuing rumors were being heard about extensions to the Temple, Rainbow and Colorow claims, northwest of the Black Dragon Mine, the Bonanza veins on the north side of the White River, and even an extension to Vernal, Utah. Mack saw the delivery that year of two more Shay geared engines, Nos. 4 and 5. Both of the Shays had been bought second hand from the Argentine Central Railway Company, and they were put to work almost immediately helping trains over Baxter Pass.

During the spring of 1911, the news was out that the Uintah Railway was going to extend its line to the Gilsonite claims northwest of Dragon. The management and clerks at the Uintah offices in Mack were busy with all kinds of plans and projects. They were not only trying to operate business in a normal manner, but they were coping with many details that came with extending the line.

Two more engines were ordered and delivered from Baldwin that year. A Consolidation, the second No. 11, was ordered to replace the aging first No. 11 on the south end of the line from Mack to Atchee. The first No. 11 was becoming expensive to maintain, and she was retired and scrapped when the new locomotive was placed in service. The second No. 11 was built to the same specifications as the No. 10, and these two engines would be the Uintah's main motive power on the south end of the line for the next 15 years. Engine 30, a large 2-8-2, was the other engine delivered that summer, and she had been ordered specifically because more power was needed for the new extension on the north end of the line. The five Shays on the roster at that time were dependable for use on the sharp curves and steep grades of Baxter Pass, but for the less-severe portion on the north end of the line, they were somewhat slow.

Along with the excitement around the two new locomotives being delivered, there also was a lot of lumber and hardware destined for the machine shop at Atchee, Colorado. The Uintah Railway was in need of a greater number of freight cars, and over the next few years, lumber, freight trucks, brake rigging and couplers all made their way across the transfer tracks to be shipped to Atchee. And along with all of this railway equipment and the normal freight traffic both going in and out of the Uintah Basin, the Uintah Railway was laying track!

Standard-gauge crossties were arriving by the car load at Mack. The ties had been ordered according to standard-gauge dimensions to save on the costs of the almost inevitable widening of the Uintah to standard gauge in the future. Along with the long ties came rail, bridge timbers, lumber and hardware to build the extension and to construct the terminal at Watson, as well as the mining camp at Rainbow, Utah. Every man that could be spared was placed in service at the transfer tracks that summer. That same summer was the only time in the Uintah's existence that there was actually a shortage of empty freight cars on the south end of the line. As the flatcars were emptied of their Gilsonite ore, they immediately were filled with supplies for the builders at the north end, and for a short few months, peak efficiency was reached as railway cars were loaded to capacity for movement in both directions.

As the Uintah Railway pushed on into the rough lands north of Dragon, the elements reminded everyone along the right-of-way about the dangers associated with railroading. The townsfolk in Mack were awakened in the predawn hours of October 1, 1911, to the yells of an accident. Just west of town, engine No. 10 had tried to cross a flood-weakened trestle and broke through, tumbling into the creek. Apprehension transfixed the residents and especially the families of the train crew and passengers on board. Everyone that could help made their way to the scene of the accident two miles west of town. The fireman and a section hand who was hitching a ride in the cab had been caught in the wreck and died.

Despite the wreck and other setbacks, the Uintah and the towns along the right-of-way were growing slowly during the next few years. The other towns, like Atchee, Dragon, Watson and Rainbow, were settling-in to become fairly permanent communities. Mack maintained a population of around 100 from 1910 to 1917, as the small railway town went about the business of transferring freight and keeping the Uintah running. The agents, secretaries and transfer gangs at the Mack terminal had a new source of traffic in January of 1914. The U.S. Post Office Department had changed the rates on parcel post shipments by increasing the size of a parcel to 50 pounds. This new size allowed shipments into and out of the Uintah Basin by parcel post to be less expensive than regular freight service.

Since the Uintah Railway had a contract to haul

the U.S. Mail in and out of Vernal, Utah, and the other communities in the basin, they were guaranteed a certain amount of money for the amount of mail they moved. The local merchants and businesses in Vernal were saving a considerable amount of money on freight shipments, the Uintah was making a good profit from its mail contract, and the federal government was paying the bill. The transfer crews at Mack saw all manner of products go from the standard-gauge D&RG to the narrow-gauge Uintah during those two years. Fifty-pound lots of lumber, cement, bricks, equipment and hardware were now being handled by the Uintah Railway instead of by truck and wagon freight. By 1916 the party was over, as businesses continually took advantage of an otherwise good thing, the government changed the regulations, changing the total pounds of goods that could be shipped or received from one location in a day.

World War I kept everyone apprehensive as the far-off conflict in Europe kept everyone talking about the concerns of Britain, Germany and France. During August of 1917, the Uintah Railway received former Florence & Cripple Creek Railroad's No. 10, a secondhand 2-8-0. The Uintah renumbered her as No. 12. She was unloaded by the transfer crew like all the other engines that had preceded her. The year 1918 proved to be a crucial year, both for America in the war and for the future of this little railway town in western Colorado. The "Highline Canal" irrigation project north of Mack had been completed during the summer of 1917, and for the first time, the desert around Mack was considered prime farm-development land.

On March 29, 1918, the federal government conducted a homestead drawing, allowing the public a chance to win the right to homestead a piece of the desert that now was under irrigation around Mack. The farmers who received the land began to relocate in Mack that spring and summer, and they greatly changed the future of the little railway town. Unlike the other small company towns on the Uintah Railway, Mack now had another economic force besides the railway to sustain it. Indeed, the farms and irrigation superstructure that were completed that summer were responsible for making Mack the only one of the Uintah's communities that would not become a ghost town with the abandonment of the Uintah Railway in 1939. The settlers, along with their equipment and other belongings showed up in Mack to make a life for themselves, and many of them came by train to the depot at Mack.

For Mack, 1918 also meant a change regarding the location of the company's upper level of management. Until 1918, the general manager and superintendent had been located at Dragon, Utah. Although they spent a lot of time at Mack, and their families usually stayed there, the Uintah provided housing for the management at Dragon. What purpose there was in their location being at Dragon is unclear. Possibly, it was to be where the "action was" concerning the production and transportation of the Gilsonite ore; however, for whatever reason, by 1918, the management was living and doing their business in Mack.

As World War I came to a close and the country got back to business, a postwar economic boom sent the demand for Gilsonite ore to record highs. Optimism was high around Mack in 1919 and 1920. The Gilsonite mines in Utah were producing at record levels, which meant there were plenty of jobs and security for the men who ran the trains and transferred the freight. Another engine, No. 40, was delivered to Mack in 1919, to help with the increased traffic on the north end of the line. She was a secondhand 2-8-2 from the New York & Bermudez Company Railroad in Venezuela. New settlers were moving to the area, as the High Line Canal changed the immediate countryside from a barren desert to fertile fields. A second general store had been built to accommodate the population that had grown from 100 to 130 from 1917–1918.

North of Mack, the farming community formed their own school district and built a school for the children in the immediate area, and it was named the New Liberty School. Transfer crews were busy with all the Gilsonite ore, and in 1920, they unloaded yet another new Shay engine, No. 6, for use on the Uintah. This brought the Uintah's total number of engines to 13. Rumors were even being heard of the Uintah extending their line again, to the Bonanza Gilsonite mines, north of the White River.

Settlers continued to tame the desert and settle in the low-lying shale hills in the countryside around Mack during 1921. However, the Uintah Railway would be changed forever by the recession that hit the country and the Gilsonite industry that year. A sudden slowdown in the shipment of Gilsonite lowered the operating income for the year, forcing the Uintah to borrow money to keep things running. Paying that debt required a large part of the Uintah's income during the years to follow. The biggest problem the Uintah faced that year was a question as to whether or not to extend the line north to new sources of Gilsonite ore. The decision was made not to build the extension due to the current economic factors, and from that time on, the Uintah Railway effectively was limited to the life of the mines at Rainbow. Compounding the problem of lower revenues from the Gilsonite freight, the Uintah already had lost the U.S. Mail contract for transportation of the mail in to the Uintah Basin in 1919. The U.S. government had started trucking the mail to Vernal that year, and it would be trucked to the basin from then on.

Because of the economics surrounding the Gilsonite industry, and for whatever personal reasons, General Manager Maurice W. (Captain) Cooley resigned his post in April of 1921. He was replaced by James E. (Major) Hood. The resignation of Captain Cooley signaled a different era for the Uintah Railway and Mack. Captain Cooley had been the general manager for the Uintah Railway since 1907. Under his leadership, the Uintah had extended its lines in 1911, expanded its locomotive roster and equipment fleet, and recorded what was one of the highest net incomes during the railway's entire history. Mr. Cooley always operated and maintained the Uintah on a grand scale, compared with many of the small narrow-gauge railways, which included a chef at the Mack Hotel, surveys for extensions, above-average maintenance and extra employees (wherever they were needed). After 1921, all of these things, except for maintenance, would be cut back to minimize expenses and the "bottom line" was the first concern. One of the first cuts in the Uintah's operating costs was the cut in passenger service between Watson and Vernal and other points in the Uintah Basin. By June of 1921, the Uintah Railway no longer offered through passenger service to points past the end-of-track at Watson, Utah. By the fall of that year, the Uintah also reduced the daily passenger service from Mack to Watson to tri-weekly service on Tuesday, Thursday and Saturday.

The 1920's saw Mack slowly growing with the ever-increasing population of farmers in the immediate area. From 1920 to 1930, the population continued to rise from 130 to just over 200. 1920 saw an addition to the back of the already overcrowded school building. However, even with the addition and the building of the New Liberty school the year before, an entirely new school building had to be constructed on the west side of town in 1923. In 1922, the U.S. Post Office changed location from the Uintah Railway's offices to the non-company house of Byron and Sarah Boyden. Mrs. Boyden was the postmistress until January of 1940, when she retired. The Boydens built an addition to the back of their house, with its own outside door during the following year, and the post office was located at that site on the south side of the D&RGW tracks as long as the Uintah operated.

By 1926, the transfer gangs, engineers and mechanics had become accustomed to periodic arrival of new motive power for the railway, and they were experienced with the unloading of a narrow-gauge steam engines off standard-gauge flatcars. This year would be different. During that summer, the men would be presented with new problems as they tried to unload what was to be one of the largest narrow-gauge locomotives to be built in the United States. No. 50 arrived from the Baldwin Locomotive Works, fresh from the erecting floor and testing track, aboard a heavy-duty flatcar. Special care was taken in the unloading of the big engine as it rolled from the flatcar, down a temporary trestle to the level of the yard at Mack. The delivery of No. 50 marked the highest achievement in motive power on the Uintah. For the rest of 1926 and into 1927, the Uintah Railway had 14 engines in service. No. 50 did exactly what she was intended to do, which was to carry more freight at lower costs than current motive power. By the summer of 1927, the Uintah was already seeing the benefits of the larger, more efficient engine, and an order for another locomotive built to nearly the same design was ordered from Baldwin. At this time in Atchee, engine No. 11 was retired and scrapped. The year 1928 saw the delivery of articulated engine No. 51 to Mack. No. 51 would be the last engine delivered to Mack, Colorado. With the efficiency this articulated locomotive provided, the transfer gangs at Mack would not have to worry about unloading engines again. Loading was another matter. That year the Uintah sold seven-year-old Shay No. 6 to the Feather River Lumber Company in California, and Consolidation No. 10 went to the Eureka–Nevada Railway. Both engines were loaded in Mack and shipped out like they came in, on the decks of standard-gauge flatcars. Meanwhile, in Atchee, Shay engines Nos. 1 and 4 were scrapped, and their good parts were saved to use for repairing the still-operating Shays, Nos. 2, 3 and 5.

By 1930, Mack was settling down to look less like a small railway town and more like a mature farming community that had a railway. The local farming community was about as big as it would ever become. The official population of the town stayed at about 200 for the next 10 years. Better roads were becoming commonplace around the country, and Mack started to see a larger amount of people on the highway running through the center of town, behind the hotel. As a result of the increase in automobile traffic, two gasoline filling stations and an automobile-repair garage were built in 1929 and 1930. By 1931, this was increased by one more filling station and a roadside cafe, and by 1935, another garage was built.

During this same period, the Uintah Railway was operating under continually decreasing schedules and service. The two articulated engines and a dwindling need for passenger service created daily mixed-train service from Mack to Watson, except on Sundays. Although for the average passenger, this was better service than the tri-weekly service that had been running since 1921, it signaled the end of any kind of grandeur that was left from the turn of the century expansionism. The Uintah no longer was a common carrier in all but the narrowest of definitions. The depot at Mack had grown accustomed to the tri-weekly schedule of the 1920's, but quickly met the new schedule of daily service with prompt attention and detail. However, by June of 1931, the Uintah again cut service to a mixed freight that ran out of Mack on Monday, Wednesday and Friday. The passenger car would then stay overnight at Watson and return on the tail end of another freight train on the following day. This schedule would stay in effect until July of 1938, when the mixed freight was cut down to two round trips per week, leaving Mack on Monday and Friday, returning from Watson on the following day. On Christmas Day of 1938, the final schedule went into effect. For the next four and a half months the depot at Mack would only see operating trains on Tuesday mornings and evenings as the train traveled from Mack to Watson and back in the same day.

Although many other towns along the Uintah's right-of-way got smaller and leaner during the 1930's, Mack stayed about the same. The Great Depression was felt just like everywhere else, but there was a sense among the railway employees that they were luckier than a lot of people. Even though there were work reductions and slow periods, there was a sense of loyalty between the management and employees of the Uintah because of the concentrated effort not to lay people off. The community grew to know and trust one another during the hard times of the Depression. Mack did make out better than many other small towns in America. Not only was there a company that was trying to make the best of it for its employees, but the local farming community was familiar with hard living, and the 1930's were just a prolonged period of the hard times that accompanied the settling of the farms just a decade earlier.

By 1939, the poplar trees that were planted in front of the Uintah Hotel and office building in Mack 34 years earlier, and the many buildings and houses that had been built since then gave the town a mature and pleasant look. Even though the town of Mack was just now being recognized as a nice community to live and work in, the reason for its very existence was getting ready to be abandoned. The Uintah Railway had filed for abandonment in August of 1938 and after a few months of public discussion and hearings the Interstate Commerce Commission approved the abandonment of the Uintah Railway. On May 16, 1939, the final mixed-freight train pulled out of Mack for the last scheduled run over the line. Combination car No. 50 was at its usual location at the end of the train, following several freight cars and engine No. 50. The small crowd of people that gathered at the depot that morning consisted of people interested in making the final run over the Uintah Railway and others who had never ridden it before. The train went to Watson that day and returned with a few more people who had used the Uintah for passenger service for the last time from their homes and businesses in the now-dying towns of Watson, Dragon and Atchee.

Gradually, non-essential workers had been laid off over the last year, and many other men had quit to find work elsewhere. Mack had quieted down considerably, as the railroaders and their families moved away. The Uintah's property was sold during the next few months, and the railway was turned over to the scrappers. The G&H Supply Company of Denver won the bid to scrap the line and by the spring of 1940 the entire Uintah Railway had been hauled off a piece at a time over the transfer tracks in the yard at Mack.

(Continued on Page 27)

FRANK A. KENNEDY PHOTO – RICHARD A. RONZIO COLLECTION

THIS SCENE was photographed in 1906, from a low hill southwest of Mack. It shows the almost complete lack of natural vegetation in this arid area of Colorado. Next to the old wooden water tank, on the left, was a material shed (with one chimney), which was located next to the Denver & Rio Grande. A Uintah section-foreman's house (with two chimneys) and company houses were near the middle. In the center of the view you can see the D&RG/URY depot. Also visible was the

Uintah Railway's office building, behind the depot. The Uintah Railway Hotel was parallel to the tracks, while the Mack mercantile building was on the far right, where the Uintah Railway trackage ended. This picture was taken as Frank Kennedy was looking toward the northeast, and it predates any construction on the south side of the tracks. Many of the railway employees boarded in the hotel, or if they had families, they lived in the company "cottages," houses that were painted white.

FRANK A. KENNEDY PHOTO – W. C. LARSON COLLECTION

THE BRAND-NEW Uintah Railway offices are shown in this picture. It was taken just after construction, and—judging by the doors leaning up against the front columns—the contractors were putting on the final touches. The two-story offices would be used by the Uintah's management to keep track of the endless supply of paper work that was created from freight shipments and passengers traveling to and from Utah. The office building was built from the same type of preformed cinder blocks as the hotel. The northeast wing of the hotel is on the right side of the scene.

FRANK A. KENNEDY, claim agent and unofficial photographer, lived in the upstairs portion of the Uintah Railway's office building. The office building contained three offices on the ground floor, Mr. Kennedy's apartment upstairs and an underground cellar. The three offices were entered by means of their own outside doors off the front porch, and the cellar door was on the west side of the building. The Uintah Railway kept its property and equipment in top shape, there would never again be such a large outlay of capital and equipment in supporting structures like the original construction in 1904, using concrete. Even the extension of the rails in 1911 to the mines at Rainbow and to Watson, Utah, would see little more than wooden structures, using the least expensive materials and hardware possible.

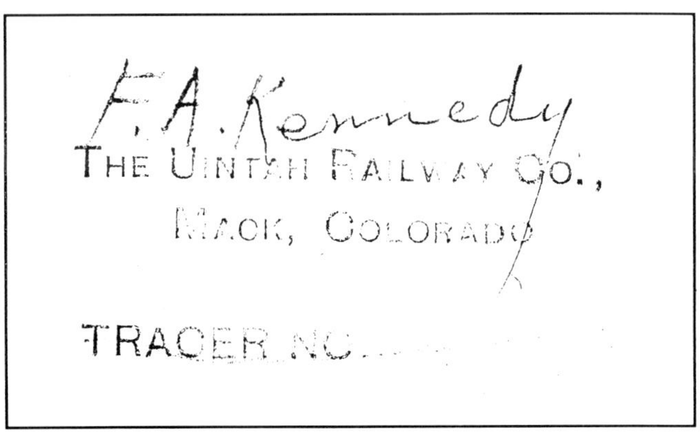

ST
CE
OFFICE
THE UINTAH RAIL
COMPANY.

FRANK A KENNEDY PHOTO – CHRIS H. UEBERROTH COLLECTION

THE FRONT PORCH of the Uintah offices was photographed around 1908. The men in the photograph are, from left to right: Arthur E. Gibson, William H. Haun, Joseph Perry (assistant secretary), Frank A. Kennedy (then claim agent and postmaster) and Haywood Brown (purchasing agent). The wire mesh was put up to allow vines to grow up the front portion of the offices. Similar terraces were also placed on the front porch and eves of the company hotel.

A LABORER TRIED to get his tools out of the way before Frank Kennedy could squeeze off a picture. In the window just above his head, an artist had lettered the glass of the post office, and you can see a selection of the photographic-style postcards on sale from Mr. Kennedy. The U.S. Post Office in Mack remained in the company's offices until 1921, when a recession in the Gilsonite industry called for cutbacks in overhead. With the changing of general managers in 1921, from Mr. Cooley to Mr. Hood, the U.S. Post Office was moved off Uintah Railway property. By 1922, the post office had been moved to the private residence of Mr. and Mrs. Boyden.

FRANK A. KENNEDY PHOTO – RICHARD A. RONZIO COLLECTION

The farming community would keep Mack growing as a town to the present day. The Mack district school closed in 1953. The business that relied on the state highway flourished during the post World War II years as traffic stopped on their way in or out of the desert in Utah. However, even that boom was relatively short-lived with the completion of Interstate 70, bypassing the town to the south. The reduction in traffic eliminated a lot of the small business prosperity that was there before. Business today is isolated to tourism and the farming community that never departed. Only a few telltale signs of the Uintah Railway can be seen. Most of the buildings that were there in 1939, have been moved or destroyed. After a fire that gutted the west wing and lobby of the hotel, the east wing has been rebuilt with a new west wall, and it has been used as a residence and business for 50 years. The old road that ran in front the hotel and offices, which curved around next to the wye, is still in the same location, as is the concrete sidewalk that was poured in front of those two buildings in 1904. Except for the "teacherage," there is nothing to be seen of the school building or the company houses. Several of the smaller private homes, located north of the transfer tracks, look much the same as they did 60 years ago.

To look at the sleepy little country town of Mack today, it is hard to imagine the activity, noise and the plumes of black smoke rising in the air, as steam locomotives went about their chores of moving freight in and out of Mack 80 years ago.

FRANK A. KENNEDY PHOTO – J. L. BOOTH COLLECTION

THE INTERIOR OF ONE of the side rooms used as an office for the Uintah Railway at Mack looked like this in December of 1908. The picture was taken as the photographer looked at the front door. Frank Kennedy, who shot the picture, had lowered the shades, and wrapping paper had been applied over the

window above the door to eliminate the bright afternoon sunlight. The men at their desks were, from left to right: Haywood Brown (purchasing agent), Joseph Perry and William Haun (clerk). Mr. Haun and Mr. Perry left the Uintah for other jobs in 1908. Haywood Brown stayed on from 1905 to 1917.

FRANK A. KENNEDY PHOTO – W. BOOTH COLLECTION

A FAVORITE LOCATION for group photographs was on the front porch of the Uintah Railway office building. Apprentice machinist, Russell Chambers, had just signed on, wearing his best suit, tie and hat for the occassion.

JOE PERRY and an Indian named "Pink Alpha" were the subjects of this picture. Why exactly this particular Indian had his photograph taken has now been lost to time, but there was always time to take pictures of Indians from the nearby Uintah reservation to help feed the appetite of the ever-curious white population. Vines were encouraged to grow in front of the Uintah Railway office to help keep the entrance cool during the hot summers on the high desert.

W.C. LARSON COLLECTION

FRANK A. KENNEDY PHOTO – DAN PRICE COLLECTION

TWO HARDY MEN and their pack animals were standing in front of the Uintah offices in this view, ready to head out toward the Bookcliff Range in search of rich mineral lodes. For the first few years of the Uintah Railway's existence, the dry desert hills crawled with men looking for places to homestead in the newly opened Uintah Indian Reservation. This was free government land for the taking, once the original Ute Indian owners had been removed through so-called "treaties." Prospectors, coal miners, ranchers, farmers and hunters would fan out into the desert and mountains beyond, in search of coal veins, spring water, oil seeps and grazing land. This swell of travelers died-out over time, as the limited resources of the area were claimed or played out.

COLORADO HISTORICAL SOCIETY – E. K. EDWARDS & SON COLLECTION

HERE IS ANOTHER VIEW of the interior, featuring a portion of the Uintah's office building in Mack. Well organized and efficient for the task for which it was built, one can make out the telephone equipment on the desk to the right. The large safe lettered "UINTAH RAILWAY" was required to store payroll funds. Railroaders commonly were paid with gold and silver coins before the practice of issuing checks was begun, as is done today.

11 11

FRANK A. KENNEDY PHOTO – RICHARD A. RONZIO COLLECTION

THE UINTAH RAILWAY'S first No. 11 was photographed in front of the Uintah offices. The original No. 11 had been built in 1880 for the Denver & Rio Grande and was their narrow-gauge engine No. 55, named the TOMICHI. She was the first engine the Uintah Railway owned. This engine was purchased during the spring of 1904, while the rails were being spiked down toward Baxter Pass and Utah. The old-style domes, oil-burning headlight, diamond smokestack and clean lines were typical of the generation when this group of engines was built by the Baldwin Locomotive Works, when ornate details and showmanship were every bit as important as function. Notice that the 2-8-0 was hauling an extra water car in the consist.

THIS VIEW OF MACK was taken from the area around the old wooden water tank that sat in the middle of the wye. Smoke from an approaching Denver & Rio Grande train can be seen on the horizon. A company house sits in the foreground just in front of the Uintah Railway Company's office building. The office building and hotel in the background were designed so similarly that they are virtually non-distinguishable from one another at this angle. The standard-gauge boxcars and outfit cars, on the right side of this picture, were sitting on a spur of the D&RG. Notice that the side-door caboose was lettered for the Rio Grande Western, the name of the route to Salt Lake City, Utah, between 1889 and 1908. (The RGW converted the narrow-gauge line between Ogden, Salt Lake City and Grand Junction to standard gauge by 1890, where it connected with both the D&RG and the Colorado Midland.)

FRANK A. KENNEDY PHOTO – CHRIS H. UEBERROTH COLLECTION

FRANK A. KENNEDY PHOTO – CHRIS H. UEBERROTH COLLECTION

THIS OVERALL VIEW was photographed from the wooden water tower at Mack, about 1915. Frank A. Kennedy had pointed his camera southeast to take this picture, providing the viewer with an idea of just how isolated Mack was during the early days, before irrigation water turned the desert into a checkerboard of small farms. The Rio Grande's mainline and the dirt road going over the horizon were both heading toward the small farming communities of Loma and Fruita, Colorado, as well as the town of Grand Junction, just a few miles away. The camera had been lined up to the left of the concrete sidewalk that paralleled the fronts of the offices and hotel. The Mack Mercantile Company and Independent Lumber Company (Coxs' store) were

at the other end of the sidewalk. This sidewalk was made of concrete in front of the hotel; however, close to the near corner, it was a boardwalk. The hotel had a large kitchen and service area in the north wing, joined together with a small hallway. The small building out back was used for storage and supplies. The frame houses in the foreground were Uintah Railway's company houses that were built in 1904, along with the hotel and offices. The small D&RG/URY depot was in-between the two railroads. Notice the "targets" on the Uintah switchstands, which indicated the direction each turnout was "lined-up for" past the wye. The train-order semaphore signal was located on the D&RG side of the depot and pertained only to Rio Grande trains.

PAUL AND BETTY GARBER COLLECTION

UNLIKE THE UINTAH HOTEL and office building, which were built of cinder blocks, the four original company houses were all wooden frame structures. These houses were just that—houses. They were much larger and spacious than section houses, and they had amenities like central heat, running water, sewer and—later on—electricity. These company-built houses were constructed to a standard plan and were the best built and most luxurious houses ever to stand in Mack. A new employee for the railway who resided in the Mack area, could only expect to get into one of these company houses after many years of attrition, through the seniority system. Once an employee moved into one of these buildings, only termination or transfer to another town on the railway could cause one of these houses to change hands. This picture of the fronts of the two southernmost houses was taken on a nice day, while several children and a dog were on the boardwalk. These wooden walkways were later replaced with concrete sidewalks. The camera had been pointed toward the northwest for this view, which included the wooden water tank that stood in the middle of the railway wye. The families who lived in these houses would not have a chance to forget for whom they worked. Day in and day out every Uintah locomotive that came into Mack had to back around the wye, literally right outside their back doors.

FRANK A. KENNEDY PHOTO – RICHARD A. RONZIO COLLECTION

AN ENLARGEMENT of a similar photograph (reproduced on Page 24) brings up details of the Denver & Rio Grande/Uintah Railway's combination passenger-freight depot in Mack. Notice the wooden platform at the depot. It also shows the Uintah Railway's office building, as well as the Uintah Hotel, perhaps as of 1910, judging by tree growth.

RICHARD A. RONZIO COLLECTION

THE ROOF OF THE TANK in the middle of the wye provided a good view of the backsides of the four houses lined up west of the office building. The white paint used on the four houses really set them off compared to the smaller drab-looking section-foreman's house, at the left. This building was built probably in 1905 or 1906. It was a smaller dwelling than the original company houses and was built close to the wye tracks and drainage ditch that ran behind town. As with all the rest of the buildings and structures that were originally built by the Uintah in 1904, this additional frame cottage was built to a smaller, more functional form. It was much like a section house, plain, simple and unrefined. Both sets of tracks for the two railroads can be seen at the right side of the picture. Coming from the depot, the Uintah's mainline split off toward the transfer track with the D&RG, while the wye switches off towards the left. The D&RG's transfer track closely paralleled the Uintah's track into the transfer yard. The D&RGW's mainline is on the far right side of the view. A large house and Crew's General Merchandise Store were adjacent to the "Ocean-to-Ocean Highway," later designated Colorado Highway 50.

FRANK A. KENNEDY PHOTO—CHRIS H. UEBERROTH COLLECTION

BY FAR THE GRANDEST BUILDING —and the largest—ever to stand in Mack, Colorado, was the company hotel of the Uintah Railway. When weary travelers stepped off the platform at the D&RGW depot, out into the desert sun, the first thing they were likely to see was the hotel, as viewed here in 1905. The hotel gave the small railway town of Mack a feeling of stature and importance. While most smaller mining railroads built good facilities for their rolling stock and depots, the Uintah spent money on public hotels and company housing for its customers and employees. The Uintah Railway Hotel, like so many other things the railway provided, was out-of-the-ordinary.

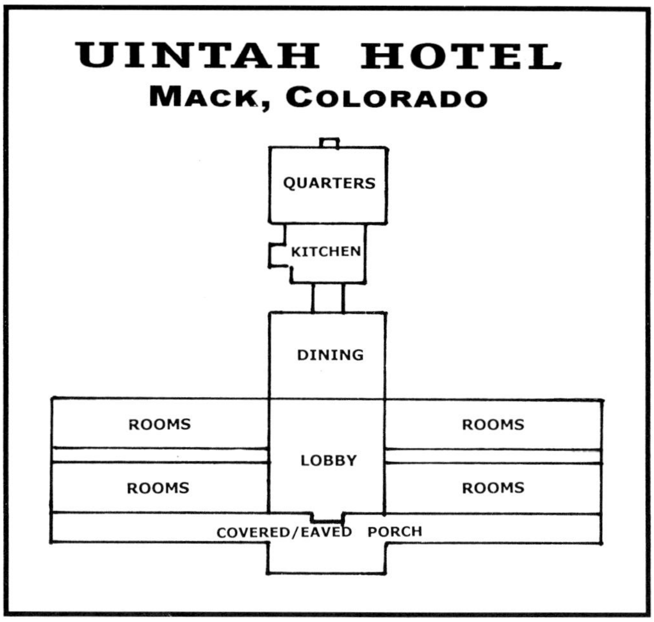

FRANK A. KENNEDY PHOTO – W. C. LARSON COLLECTION

THE LOBBY AND ENTRY WAY into the hotel shows the magnitude and luxury of the establishment. Wooden floors, large rugs, sturdy furniture, light fixtures, paintings and taxidermy all did their part in making the guests feel like they were somewhere special. The doors in the back went into the dining room. Beyond the dining room were the kitchen and staff residence.

FRANK A. KENNEDY PHOTO
COLORADO RAILROAD MUSEUM COLLECTION

LOCATED JUST BEHIND the main lobby in the hotel was a large dining hall used by the Uintah to feed the guests of the hotel and others in town who needed a place to eat. Whether it was for the continual stream of travelers on their way to and from the Uintah Basin, or Christmas dinners, or for large groups of construction workers for the D&RGW, the Uintah Railway Hotel prided itself on serving large meals, good food and manners that were a step above the usual boarding house. In this picture the tables were "prepped" and ready for the next round of hungry railroaders and travelers. Electric lights illuminated the darkness from chandeliers.

FRANK A. KENNEDY PHOTO
COLORADO RAILROAD MUSEUM COLLECTION

WHEN UPPER-LEVEL management and stockholders from back East would come to survey the Uintah mining district with unsophisticated wonderment and investigate the Gilsonite fields in eastern Utah, they would be afforded a comfortable atmosphere in which to stay. The entry hall for the Uintah Railway Hotel was decorated like a hunting lodge. This was a popular style at the turn of the 20th century, especially for "getaway" resorts and hotels "way out West." This large room, with its open-frame ceilings, was used to impress guests and entertain crowds for years to come. Rooms were accessible through the draped doorway.

RODGER POLLEY POSTCARD COLLECTION

THE MACK DEPOT, at the left, was shared by both the Denver & Rio Grande and Uintah Railway. It was built across the road from the U.S. Post Office and other Uintah Railway offices and residences. The row of poplar trees planted at curbside and the broad-leaf trees that lined the opposite side of the sidewalk, along with the lawn that surrounded them, produced an inviting atmosphere for local residents and travelers alike.

THESE YOUNG LADIES were wearing fur warming muffs to protect their hands from the chilly air of Colorado's high western desert. The lack of insulation under the roof of the hotel is readily apparent in this winter illustration—from the way most of the snow had melted off the roof.

FRANK A. KENNEDY PHOTO – W. L. BOOTH COLLECTION

FRANK A. KENNEDY PHOTO – CHRIS H. UEBERROTH COLLECTION

IN THIS ATTRACTIVE SCENE, the late-day sun cast long shadows across the lawn in front of the hotel. The small lined ditch along the row of poplar trees provided water for irrigation from up in the Grand Valley. A screened area on the porch provided a ventilated area to sit and enjoy the cool evening air.

FOR SEVERAL YEARS, the Uintah Hotel was known for being a cool retreat for holding social gatherings and community events in the hot, arid climate of far-western Colorado. This picture was taken facing the southeast wing of the big hotel, where guests stayed in their rooms.

FRANK A. KENNEDY PHOTO – GEORGE ROSELAND COLLECTION

RODGER POLLEY COLLECTION

THIS PRE-1920 VIEW of the Uintah Hotel's dining room shows a festive atmosphere at one of the many ladies' luncheons hosted by the popular Daughters of the American Revolution. The ladies in the photograph were, from left to right: Fay Waggoner, Mrs. Captain Cooley, Opal Hunt, Mrs. Ainsworth, Peggy McElvaney, Mrs. George Lyman and Mrs. Fitzpatrick. Mrs. Cooley was the wife of the general manager of the Uintah Hotel, M. W. "Captain" Cooley. The Cooleys originally were from England, and the railway hotel was known for having a rather prim-and-proper atmosphere. No matter how hot the summer days were, the "Captain" would always put on a dinner jacket and tie for the meals being served in the dining hall. The hotel's kitchen was accessible through the swinging doors behind the table.

BILL AND SUE KNOUS COLLECTION

FROM LEFT TO RIGHT: Superintendent Walter Rader, Claim Agent Frank Kennedy and General Manager Lucian Sprague, posed on the lawn southeast of the Uintah Railway offices during 1926. The photographer also included the row of company houses in the background, at the right.

THESE COMPANY PHOTOGRAPHS portray the Uintah Railway's fine dining china.

CHRIS H. UEBERROTH COLLECTION

THE DAUGHTERS of the American Revolution was a popular social club, and many of the ladies in the Grand Junction area were members. Mrs. Cooley, wife of the general manager of the Uintah, was a supporter of the DAR and frequently hosted events at the Uintah Railway Hotel. These pictures show one of the gatherings that took place during the summer of 1913, photographed in front of the Uintah offices. On this occasion, several of the ladies were given a special trip on the Uintah to Columbine for a picnic. Mr. Falvey, then the superintendent of the Uintah, operated Railcar 50, taking Mrs. Cooley and six other women from Grand Junction to the cool alpine meadows on the north side of Baxter Pass. The Grand Junction newspapers reported several other trips similar to this one over the next few years. On one occasion, they only made it to Atchee and back, while on another trip, they had their picnic as far north as McAndrews Lake. Whatever the distance of their trip, there would always be a well-prepared dinner at the Uintah Railway Hotel waiting for their return. During this time, the Uintah Railway Hotel often was described as the Mission Hotel. No other reference to this name has been found.

DAR COLLECTION – MUSEUM OF WESTERN COLORADO

RODGER POLLEY POSTCARD COLLECTION

MR. W. H. COX stood proudly in front of his establishment in Mack about 1914. This store was one of the first buildings to be built in Mack that was not owned by the Uintah Railway. Just southeast of the Uintah Railway Hotel, the store was a short walk from most of the residences in town, located at the southeast end of the Uintah's track. The original building was built in 1904 and consisted of only the part just behind the false front. That building was operated as the "Fruita Mercantile Company" by a Mr. Richard Stewart. This name came from Fruita, Colorado, nine miles to the east, and probably was a branch of the Fruita Mercantile that was located there. Mr. and Mrs. William and Jennie Cox took over the business around 1909, and it was named the "Mack Supply Company." Later, the name was changed to "W. H. Cox & Company, General Merchandise," and by 1913, Mr. Cox had gone into business with the Independent Lumber Company out of Grand Junction. It was named accordingly, and a much larger complex was constructed around the original building; a large lumber warehouse was built across the D&RGW tracks to the south. This picture of the west side of the building shows the extent to which Mr. Cox went to increase the size and appearance of his establishment. Mr. Cox was standing on the platform of a large scale that was set into the ground to enable an entire wagon to be weighed. The screened-in section was added to the left in front of an addition to the earlier building. At the east end of the building a larger two-story structure was built and several large out buildings were built to the south. All of this was fenced off with a tall picket fence and then white washed to give a fresh new finish to the business. By 1916, Mr. Cox was bought out by the Independent Lumber Company, which for a number of years, operated the merchandise and grocery story separately from the lumber and hardware store on the south side of the tracks. By 1924, the two buildings separated by the D&RGW tracks were both named and operated by The Independent Lumber Company. This store was managed by a Mr. Clem Miller into the late 1930's.

DURING ITS FIRST 20 YEARS of business, the Mack Mercantile building carried on a brisk business. This picture of the back side of the building was taken before 1910 with a large group of mules ready to pack provisions for a group of men going north into the Bookcliff Mountains, or west into the desert. All manner of goods and packages can be seen strapped to the mules' backs. Two bales of hay, a box of canned "Champion Peas" and a two-burner camp stove were among the many sacks and boxes that made up this "mule train." Isolated ranches, farms and mines, along with settlers, prospectors and surveying parties kept the little general store at Mack busy as the "jumping-off point" for people living in the largely uncivilized and non-populated area to the northwest. With the continued growth in population of the area, there was a steady stream of produce and materials needed every day. As irrigation was provided to the desert area north of Mack, local farms sprang up all around Mack, and each one needed a lumber and hardware store to fill their needs. During this same time period, the Uintah Railway was continually growing, also. The 1911 extension to Rainbow and Watson, Utah, and a constant growth in equipment and employees meant only good times for the Mack Mercantile. Whether it was lumber for a small house being built in Atchee, or burlap sacks and string for the mines in Utah, or simply some canned food for the section men at Baxter Pass, the Mack Mercantile and Independent Lumber Company had a steady stream of orders for the "morning train." This photograph also shows a good view of the south side of the original one-room schoolhouse that was built by the Uintah for children in Mack.

RICHARD A. RONZIO COLLECTION

MUSEUM OF WESTERN COLORADO – RESEARCH CENTER – BOWEN COLLECTION

DURING THE "BOOM DAYS" in the early teens, the Mack Mercantile & Independent Lumber Company was doing enough business to stand the competition of a rival grocery store opening its doors. The small wooden structure south of the D&RGW tracks was originally just a small house that was turned into a grocery store with the addition of a fake front. It is unknown who first used it as a store, but it shows on the 1915 ICC maps. By 1918, it was known as "Crews General Merchandise," owned and operated by a Mr. Luke J. Crews, who was previously a work-gang foreman for the Uintah Railway. Owing to a lackluster performance on the part of the Independent Lumber Company people to run the grocery portion of their business, Mr. Crews saw an opportunity to get in on the local grocery business. Crews Grocery was a fixture in Mack into the 1930's. This early photograph of the east side store front was taken during the early days of Mack, before much vegetation was evident in the area. In the far distance, a standard-gauge boxcar was sitting on the transfer track, waiting to be filled with heavy bags of Gilsonite ore.

RODGER POLLEY COLLECTION

THIS SCENE WAS SHOT from the tail end of the wye, looking northwest toward the "business district" in Mack during the early 1930's. Mesa County had started maintaining the road from Grand Junction and Fruita to Mack, and on to the Utah border, in the mid-teens. The small white building and the Sinclair "filling station" were two of three different gasoline stations in Mack during the 1930's. The large peaked-roof building was the Mesa County District School. The Mack Cafe was the oldest of the buildings in this part of town and served more than just food. During prohibition years, this was the only place in town that had any alcohol available for consumption. Typical of "prohibition America," everyone knew that alcohol and an occasional game of poker could be found in the back rooms of the Mack Cafe, but nothing was ever done about it.

EVERY NOW AND THEN, someone would turn in the owners and patrons to the county sheriff or city police in nearby Fruita. Local law-enforcement officers would promptly call the Mack Cafe and tell them that there was a complaint and that they were coming to inspect the property. A concerned citizen would be posted at the front door looking east up the highway for several miles to the point where the highway crested a small rise. As the headlights of the approaching police car topped the hill, everyone was alerted to put the drinks in the back room, and the patrons who were not in the best condition, were "shown the way home." When the law arrived, there was no occasion to search any further than the front room because there was no trace of illegal alcohol or its detrimental effects to be seen.

RODGER POLLEY COLLECTION

THE MACK SERVICE STATION and Garage was operated during the early 1930's by a Mr. George Davis. This station consisted of a gasoline fuel pump and a single-car garage, with small living quarters behind it, and it was thought to be located on the west end of town. However, the Uintah flatcars, in the background, show that this station was not far from the rails. It is hard to pin a name to any one location because, during the Depression era, as fast as someone put up a sign, they seemed to go out of business, or sold it to someone else. Some of the names of these small service station were as follows: Mack First and Last Chance Filling Station, Mack Service Station and Garage, Shell Filling Station, Sinclair Station and State Line Store.

THE U.S. POST OFFICE in Mack was originally in the Uintah's office building, and for a time, it was reportedly in both the Uintah Railway Hotel and the D&RGW depot. By 1922, the Mack post office had been moved to the Boyden's house on the south side of the tracks. Mrs. Sarah Boyden was the postmistress from 1922 to 1939. Mr. Boyden was a part-time rancher and lawyer. Together, they operated the post office, first in their house, and then in an addition off the back side of their house. This little post office was at the center of everyday life in Mack, Colorado, during the 1920's and 1930's. Correspondence from family and friends was always looked forward to in the era before telephones became commonplace in every home and business.

GEORGE ROSELAND COLLECTION

RODGER POLLEY COLLECTION

THIS BOARD-AND-BATTEN building was constructed during the teens as a small saloon and barber shop, across the road from the Uintah's wye. It was advertised as a billiards shop during the prohibition years, and during the 1930's, it was operated as the Mack Cafe. The main part of the building was a large hall that—over time—saw barbers' chairs, a small bar and a pool table replaced by chairs and tables. Various owners and managers lived in the small quarters at the back of the building. In this picture, taken about 1920, a few trees had been planted in front of the building, and posts and lumber have been erected to protect the trees from the children, horses and automobiles.

THE FIRST SCHOOL BUILDING in Mack originally was the Uintah Railway's yard-office structure, built by the Uintah between the tracks of the URY and D&RGW. After the first school year in 1906–1907, the building was moved from this unsightly and dangerous location to a lot northeast of the Mack Mercantile building. The small one-room school building was used until 1920, when a second small classroom was added to the back of the building. By 1923, the building in this view was then moved or torn down. The Mack Mercantile store was at the right, while a group of mules and a single horse waited for their master to lead them on a journey to re-supply some far-off ranch or mining claim.

GEORGE ROSELAND COLLECTION

BETTY COSSEBOOM COLLECTION

A GROUP OF SOME of the first students and teachers were standing in front of the entrance to the schoolhouse in the 1913–1914 school year. Cleone Fitzpatrick is the girl in front of the door, with the knee-high dress. She was the only one who was positively identified in the picture, but it is known that one of the lades in the doorway was the teacher, Harriet Weir. The two boys were: Leonard and Earl Cox, and the four girls were: Josie Cox and three sisters, Marguerite, Bertha and Lucille Marshall.

W. C. LARSON COLLECTION

THE LAST MACK SCHOOL was a much nicer structure than the school building it replaced. School District No. 44 was created in 1922 to raise tax money to build and support a new school. Until then, the Uintah Railway had rented the one-room building to the school for $1.00 per year. With the increasing costs associated with a larger student body, as well as an increasingly cramped building, the Uintah no longer wanted to be directly involved in the business of schools. The new school building was built during the autumn and winter of 1922–1923, and occupied during the summer of 1924. Two large classrooms divided the building in the rear, while a large auditorium occupied the front portion of the building. The gate in the front of the photograph was used by most of the students to enter the property, and it was the scene of many practical jokes. It seems that this location was a natural place for the children to come running through when the bell was ringing. Naturally, more than a few different menacing boys figured that this was a good location for a trip wire!

THE NEW SCHOOL at Mack was used for 30 years, from 1923 to 1953. With the abandonment of the Uintah Railway in 1939, the class sizes quickly became smaller. As time passed, the use of the school was diminished, as most of the students in the area were either closer to the New Liberty country school or to Loma. This view was taken on the front steps of the school during its prime. Another large group of children posed on the merry-go-round that was on the west side of the building.

W. C. LARSON COLLECTION

RODGER POLLEY COLLECTION

WELL-BEHAVED KIDS—*for a few minutes anyhow!*—**posed at the new Mack school entrance. Just inside the doors were coat racks, and a place beneath a bench was provided for overshoes. And the pupils remembered that the teacher had a wooden ruler to maintain discipline!**

W. C. LARSON COLLECTION

BYRON BOYDEN was the promoter of the Uintah golf course and can be seen here taking a swing from the No. 1 tee box. The course was on the hills north of town, with the tee box located just through the front gate. The large steel water tank, at the end of the wye, is seen here along with the large Gilsonite ore house that is in the far background. This "country club" at Mack was a social organization that started in the early 1920's to provide activities for the townspeople. Dinners at the Uintah Railway Hotel and dances at the newly constructed school building raised money for building the tennis courts, and, north of town, a golf course. The nine-hole course was one of the first golf courses built in Mesa County, and it became somewhat of a popular destination point for avid golfers. Over time, better courses in Fruita and Grand Junction pulled the participants away, and by 1938, the course was closed.

A YOUNG LADY, who was a baseball player, appeared to be ready for the game to commence. Not to be outdone by communities like Fruita, Loma and even Grand Junction, the residents in Mack were always participating in some kind of sport or game. The Uintah Railway had its own baseball team, which traveled to the other local communities on a regular basis. This became a big pastime for the residents of small railroad towns like Mack, Atchee and Dragon City. Not only did they provide players for the Uintah's team, but they also developed fierce competition between these small towns. It had even been said that how good one was at baseball could have an effect on if one got hired by the railroad.

THORNE COLLECTION

RODGER POLLEY COLLECTION

AT THE TURN of the century, the United States was being developed and its natural resources tapped. A large migration of immigrants flowed into the country and, by necessity, found their way into construction, mines, and transportation. Everywhere there was work to be done. The Uintah's original group of employees had a large group of Greek immigrants who found jobs not only as laborers for the Uintah, but also at the Gilsonite mines in Utah. This picture of a group of Greek employees was taken in front of the section house that was west of the wye. Several employees and their families that made up the Mack to Atchee section gang stayed in this house. The picture was taken looking south, with the large wooden water tank in the background and a D&RG material shed beyond the mainlines of both railroads. The small log building that is set into the ground was the company's first ice house.

WITH THE GREAT DEPRESSION of the 1930's, a hard time for many Americans, was just another fact of life. Many employees of the Uintah Railway and many of their children remember General Manager Walter Rader as a thoughtful man and boss. While numerous industries, including the railroads, laid off their work forces in large numbers, Mr. Rader was well known by his employees as going to great lengths to maintain jobs for as many people as possible. Instead of layoffs, men were cut back to three-day or four-day schedules. Although this hurt the pocketbook, it still allowed a man to bring home something. Reminders of how bad things could get in the Depression were all around. The U.S. Government initiated the Civilian Conservation Corps (CCC) at the behest of President Franklin D. Roosevelt, in the 1930's—to enable men to work at something for a little pay. In the arid region around Mack, this meant building thousands of small earth dams in the washes and gullies to hold back runoff water. This picture of a CCC camp just west of Mack was taken in 1937. As can be seen by the row of canvas army tents, tough living conditions and hard work were a way of life for men who were "down on their luck."

WOODY BOOTH COLLECTION

FRANK A. KENNEDY PHOTO – RICHARD A. RONZIO COLLECTION

A PERSON COULD SAY that the Uintah Railway's whole reason for existence was visible in this picture of the transfer track at Mack. The steady flow of Gilsonite from the mines in Utah was the life blood of this railway, and getting the ore to standard-gauge cars of the Denver & Rio Grande Western Railroad was the line's biggest task. In this view of the transfer yard, you are facing the southeast—in the direction of the D&RGW/URY depot. At the mines in Utah, the miners sacked and marked the Gilsonite ore according to the specific qualities found in a certain section of the mine. It was then shipped via the Uintah Railway to a specific car destined to the industry that had placed an order. The bags usually weighed close to 200 pounds each, and it required two men to maneuver one bag at a time into standard-gauge boxcars. Inside the boxcars, the large bags had to be lifted and stacked into tiers. It was said the work inside the cars was either unbearably cold or suffocatingly hot, but nowhere in-between. You can see a

string of Uintah flatcars, loaded with Gilsonite sacks, in the background. From left to right, these laborers were: Chris Pappas, L. B. Higgins, a Mr. Perkins, Herschel Shook, a Mr. Walters Sr., Jim Booth, Frank Wagaman, Paul Calderwood and Dean Watters. The year was 1925, and Uintah Engine 10 was sitting under the water spout as water was being delivered from the wooden three-tank water car, next to the water tank and boilerhouse. The roof of the icehouse was in the background.

PAUL AND BETTY GARBER COLLECTION

ON THE 1915 MAP of Mack, this small tar-papered shack was listed as a bunkhouse and was northwest of the wye, on the north side of the tracks. While the trainmen and their families lived in the nicer company houses near the Uintah offices, the laborers and section men lived in less comfort. This early photograph of this small bunkhouse shows a group of Greek immigrants posing for the camera in front of the place they called home. You may wonder why all of them had "dressed up." Was it for the photographer's sake, or was something taking place with the table, papers and ink well? To the left of the men in the shade, a dog was taking a much more relaxed attitude toward the photographer. Over the dog's head, in the distance, the Uintah's wooden water tank stood in the middle of the wye.

UINTAH TRANSFER WORKERS were photographed in the yard at Mack, ca. 1915. The Uintah Railway originally utilized a large wooden water tank, which had been built above a tall enclosed pumphouse, situated inside the wye. The water tank was an unusual building, and it was one of the largest railroad structures on the entire line—other than the enginehouse at Atchee. The lower pumphouse was built in a six-sided hexagon pattern, with a door or window on each side. The massive timbers that held all the water's weight above it were directly below the tank. John Eagan, a foreman for the Uintah was nearly killed during the 1920's, when these timbers finally gave way, crushing the pumphouse and nearby buildings, giving the transfer tracks an unexpected bath. Mr. Eagan had been working in the pumphouse just minutes before the accident, and while leaving the premises, he was washed off his feet and suffered minor injuries from the debris.

RODGER POLLEY COLLECTION

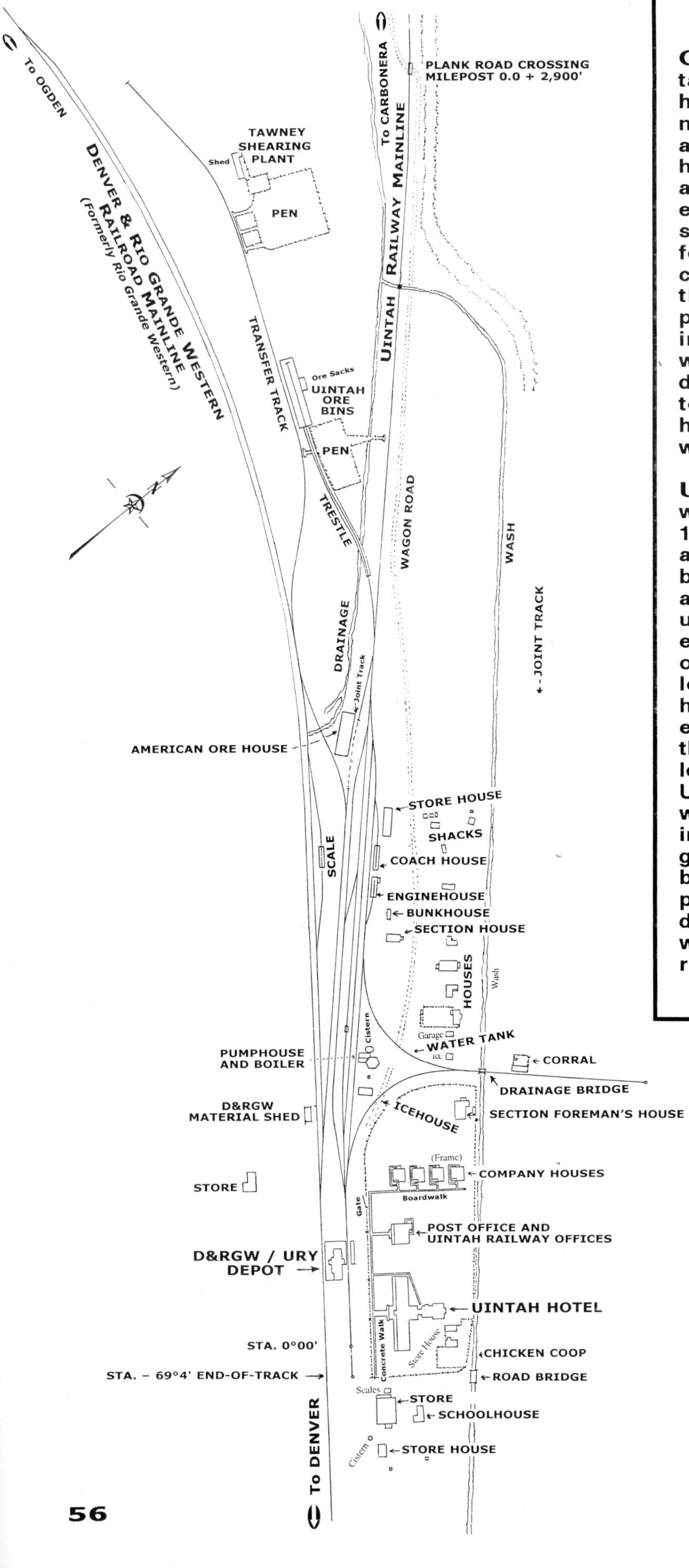

UINTAH RAILWAY
MACK, COLORADO
— Circa 1915 —
By Rodger Polley

0' 100' 200' 300' 400' 500'
SCALE IN FEET

BELOW: This group of workers was standing on top of a structure known only as the Uintah ore bins on the 1915 map. There is evidence to suggest that the Gilsonite mines in Utah hauled their ore out in loose form for a few years, rather than in the standard sacks. At the northwest end of the railroad terminal, the Uintah's trackage climbed up out of the yard on a long curving trestle, which was used to take the gondolas to the top of the bins. Drop-bottom doors were opened on the gondolas, allowing the Gilsonite ore to fall into the bins, awaiting transfer to standard-gauge cars on the south side of the bins. It should be noted that other than this picture and a 1915 map, there is no additional evidence to support the theory that what you see below the workers is Gilsonite. The Black Dragon mine did have a large ore bin or hopper-looking arrangement in front of it, and they did use small dump-style cars at that mine. Basically, the Black Dragon was the only mine producing any large quantities of ore before the extension to Watson and Rainbow in 1911. In any case, the Gilsonite mines and industries they supplied became increasingly choosy about the grades of ore that were coming out of the mines, and eventually, market forces and the increased distance in the mines from the ore face to the rail head conspired to end all transportation of Gilsonite in loose form. In the 1915 map, an "ore sacks house" is located on the north side of the bins, suggesting that the marketplace preferred sacked Gilsonite ore.

PAUL AND BETTY BOOTH GARBER COLLECTION

FRANK A. KENNEDY PHOTO – DR. LESTER WILLIAMS COLLECTION

THIS WAS A VIEW of the transfer yard in 1925, as you looked northwest from the top of a flatcar loaded with Gilsonite ore. The roof for the Gilsonite-ore warehouse is in sight above the boxcar tops, while the American ore-house roof was beyond the boxcars. The Uintah's water tank and icehouse were to the right, and the boiler-house (with the smokestack) was on the other side. Railcar No. 50 was sitting on the other side of the water tank, on a small spur next to the wye. Beyond the small railcar stood the single-stall enginehouse, used to house one of the Uintah's engines when not in service. The Uintah's loaded flatcars were lined up on the parallel track, next to the standard-gauge boxcars on the D&RGW.

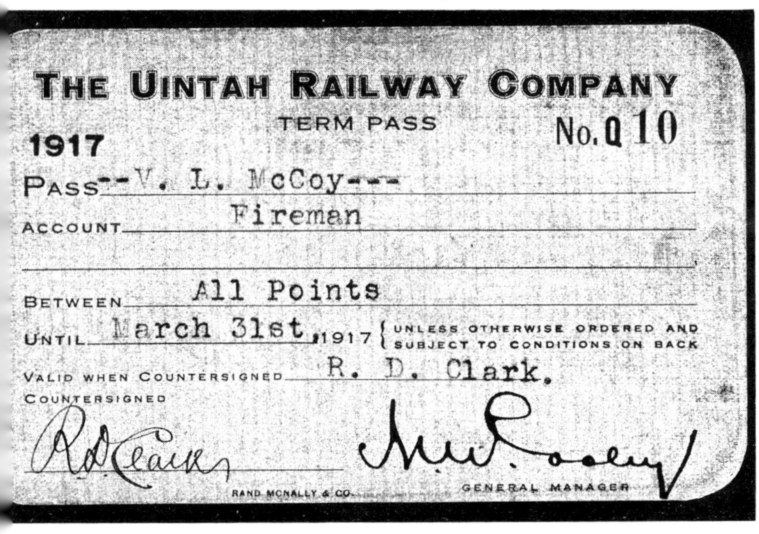

THE UINTAH RAILWAY COMPANY

TERM PASS

1917 No. Q 10

PASS --V. L. McCoy---

ACCOUNT Fireman

BETWEEN All Points

UNTIL March 31st, 1917 { UNLESS OTHERWISE ORDERED AND SUBJECT TO CONDITIONS ON BACK

VALID WHEN COUNTERSIGNED R. D. Clark.

COUNTERSIGNED

R.D. Clark M. W. Cooley

RAND MCNALLY & CO. GENERAL MANAGER

RICHARD A. RONZIO COLLECTION

FRANK A. KENNEDY PHOTO – RODGER POLLEY COLLECTION

AN EXCAVATION had been completed for the scale track at Mack. The cut at the left was for the scale-control mechanism. Freshly poured concrete was being tamped down with a chunk of 6 x 6-foot timber, as men on the bank prepared to wheel more mix down into the excavation—one wheelbarrow at a time. Several workers were mixing and pouring concrete by hand for this excavation. The section-foreman's house (with two chimneys) is visible in the distance, near the tail track of the Uintah's wye. This photograph relates to the map on Page 56.

THE UINTAH RAILWAY'S FIRE SERVICE FOR MACK...

Although there are no recorded fire incidents in Mack, there was a plan of attack, if there was any problem. A company memo from 1910 was distributed around Mack, and—like employees in the other towns along the right-of-way—everyone was responsible for doing his part if a fire was to break out:

Fire plugs are located as follows:

#1 Plug - a long side stand pipe opposite warehouse
#2 Plug - in front of school house
#3 Plug - in rear of cottage occupied by Mr. Gibson
#4 Plug & #5 Plug occupied in corridor of hotel
#6 Plug - in pantry of hotel

Keys for the fire plugs are stored underneath water tank

Fire Hose:

4 - 50' lengths of hose with nozzles are stored underneath water tank

3 - 30' length of hose in pantry of hotel. Fire hose stored under water tank has different threads than hose in pantry of hotel. If necessary to connect the hoses a coupler for such purposes are stored in the room under the water tank house. In case of fire about warehouse, auto house, or section house, hose connection should be made with plugs #1 and #2. In case of fire about school house or cottages, hose connections should be made to plugs #1 and #2. In case of fire at hotel, office or depot, hose will be found connected to plugs #4-5-6 in hotel building and hose found under water-tank house should be connected to fire plug #2.

EMPLOYEES WILL MAKE THEMSELVES FAMILIAR WITH ABOVE INSTRUCTIONS AND BE GOVERNED ACCORDINGLY.

E. A. Groves, Superintendent

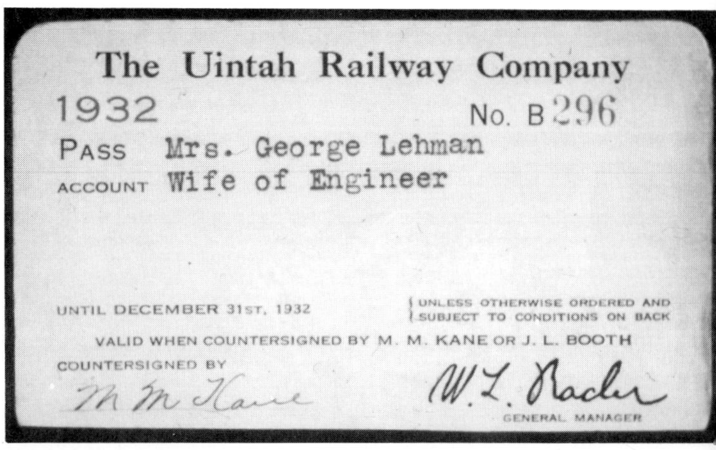

The Uintah Railway Company

1932 No. B296

PASS Mrs. George Lehman

ACCOUNT Wife of Engineer

UNTIL DECEMBER 31ST, 1932 — UNLESS OTHERWISE ORDERED AND SUBJECT TO CONDITIONS ON BACK

VALID WHEN COUNTERSIGNED BY M. M. KANE OR J. L. BOOTH

COUNTERSIGNED BY M M Kane

W. L. Rader
GENERAL MANAGER

RICHARD A. RONZIO COLLECTION

INTERSTATE COMMERCE COMMISSION

— June 30, 1919 —

Inventory of the Uintah Railway MACK, COLORADO

STATION AND OFFICE BUILDINGS

–Depot, a combination passenger and freight station
 –Frame wood, shingle roof, lighting and foundation platforms.
 –Jointly owned by D&RG Ry. and Uintah Ry. 50% : 50%
–Express building
 –Frame wood, composition roof, foundation
 –16' x 18' x 12'
 –Jointly owned by D&RG Ry, and Uintah Ry. 50% : 50%
–General office building
 –Concrete block, shingle roof, porches, heating, plumbing, lighting.
–Agent's house
 –Frame, wood-shingle roof,
 –porch, plumbing, lighting, foundation.
–Ice House
 –Frame, composition roof
 –27' x 48' x 21.5'
–100-ton track scale, 50'
–Two stockpens

ROADWAY BUILDINGS

–Section Foreman's house
 –Frame, wood shingle roof
 –16' x 30' x 12' + lean-to 8' x'8' x 9.5'
–Material shed
 –Frame, galvanized iron roof
 –24' x 66' x 12'
–Bridge & Building storehouse
 –Frame, corrugated galvanized iron roof
 –12' x 60' x 15.5'
–Tool house
 –Frame, composition roof
 –12'x 46' x 12'
 –Lean-to 4' x 12' x 9.5'

WATER STATIONS

–22' diameter tank, 16' high on 30' high timber frame
 –concrete foundation
–10,000-gallon steel cistern
 –pump, pipes and water stand

FUEL BUILDINGS

–Coaling-dock station, 12' x 36'

SHOPS & HOUSES

–Shop building
 –Frame, wood shingle roof
 –Addition 12.5' x 16' x 10.5'
 –Lean-to 10' x 22.5' x 9.5'
–Frame houses
 –Frame, wood shingles,
 –foundation, lighting, plumbing
 –Superintendent's house
 –Yardmaster's house
 –Engineer's house
 –Residence
 –Conductor's house

TELEGRAPH & TELEPHONE LINES

–Various office apparatus
–80 cells of batteries

POWER PLANT BUILDINGS

–Electrical distribution and yard lighting

SHOP MACHINERY

–13' long 34" diameter locomotive-type steam boiler
 –Salt Lake Hardware Company, 1903
–7" x 12" steam engine
 –Erie City Iron Works
–Fairbanks-Morse Electrical generator, 115 volts
–Air compressor and two air reservoirs

UINTAH RAILWAY

MACK, COLORADO

— Circa 1930 —

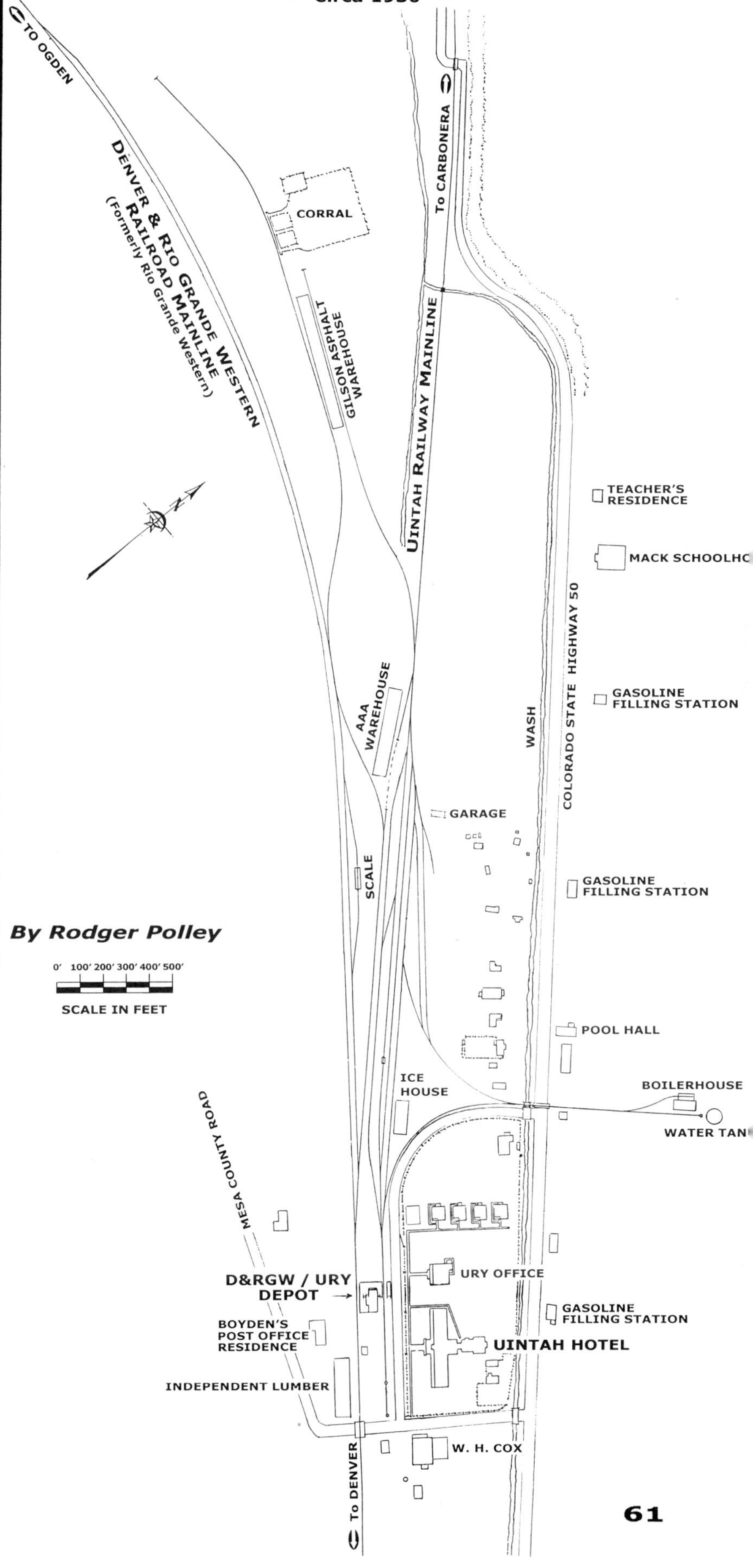

FRANK A. KENNEDY PHOTO – RICHARD A. RONZIO COLLECTION

THIS UNUSUAL CABOOSE was used on the Uintah for only a short while in 1907. Perhaps rebuilt from the railbus on Page 61, it remarkably looked like a four-wheel British "guards van" (used on British freight trains). Probably, it was not a long-lived piece of equipment just by the way it looked. A lack of any kind of stove or heat source, coupled by the many windows, would have made for a cold ride in the winter. The single-pane glass in the windows would prove to be less durable than tongue-and-groove lumber; this picture shows one window already cracked and broken. The train of drop-bottom gondolas, with tarpaulins on them, is another mystery. During the early years, the Uintah used tarpaulins thrown on top of loads of Gilsonite ore to lessen the chances of fire. A spark from the engine could easily ignite a car loaded with the highly combustible material. At different times during the mining of the Black Dragon mine in Utah, management decided to try to haul loose Gilsonite. As a result, the Uintah transported a certain amount of Gilsonite ore by gondola. This was one of those few trains of this type. The tarpaulins placed over the Gilsonite ore would be useful, not only for protecting it from an accidental fire, but also to keep the finer particles from blowing away. Most of the Gilsonite ore was bagged in burlap sacks at the mine and loaded by hand, as can be seen by the large stockpile of sacks behind the caboose. The front of the single-stall enginehouse is visible, along with a water column and the track leading to the wye, off the right side of this picture.

DURING THE LATE TEENS, the Uintah demolished the ore bins and erected a large Gilsonite storage warehouse in its place. As uses for Gilsonite grew, orders were placed by the various industries for specific types of ore. Depending on how it was formed in the mines, Gilsonite could have an array of different chemical and physical qualities. As each bag was taken from the mine, it was designated and tagged accordingly. When these bags reached Mack, they were sorted in this warehouse for shipment in standard-gauge cars for the trip to big-industry users in the East. Although a great majority of the Gilsonite was transferred directly from the Uintah's narrow-gauge cars to standard-gauge cars at the parallel tracks in the yard, a large quantity of the bags were also stored and later transferred through this warehouse. In this picture, you are looking at the north side of the warehouse from the Uintah's mainline, with the tracks full of flatcars and boxcars of Gilsonite ore, waiting to be unloaded. On the south side of the warehouse, a string of standard-gauge boxcars were waiting to be filled.

PAUL AND BETTY BOOTH GARBER COLLECTION

FRANK A. KENNEDY PHOTO – RICHARD H. KINDIG COLLECTION

UINTAH RAILBUS No. 50, as completed in 1905, was equipped with a six-cylinder gasoline motor, with a four-wheel rear truck providing the propulsion. General Manager Cooley and other Uintah executives used this railcar to traverse the line. The three-seat vehicle was capable of carrying six adults, and it is shown here at the depot platform in Mack.

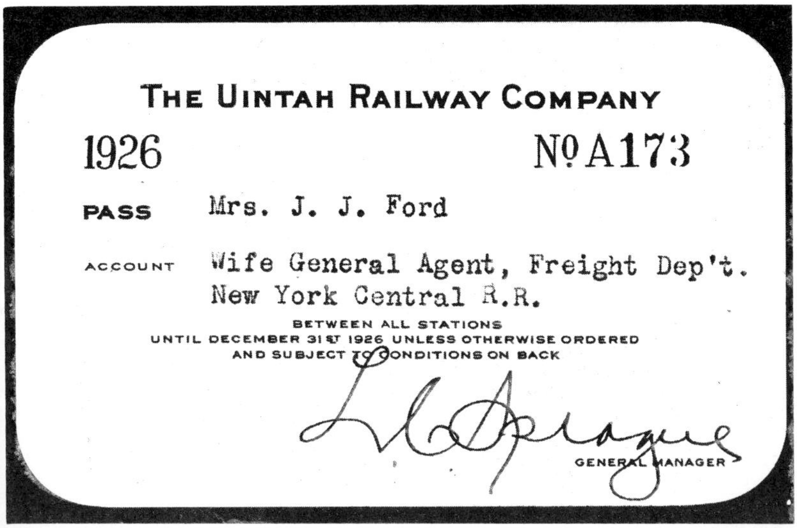

THE UINTAH RAILWAY COMPANY

1926 №A173

PASS Mrs. J. J. Ford

ACCOUNT Wife General Agent, Freight Dep't. New York Central R.R.

BETWEEN ALL STATIONS
UNTIL DECEMBER 31ST 1926 UNLESS OTHERWISE ORDERED
AND SUBJECT TO CONDITIONS ON BACK

GENERAL MANAGER

RICHARD A. RONZIO COLLECTION

ED AND NANCY BATHKE COLLECTION

THE HINGED DOOR at the front of this unusual European-style railbus hid the number of the vehicle—although it looks very much like a different version of No. 31. In any case, the railcar was ready for departure at the Mack depot. Possibly, this machine was powered by a gasoline motor. The cab and passenger compartment contained 13 windows on each side, including door glass for six doors. This railbus may have operated on the Uintah until 1910.

FRANK A. KENNEDY PHOTO – RICHARD H. KINDIG COLLECTION

THIS STRANGE-LOOKING passenger-carrying vehicle was lettered "UINTAH 31." The railbus was powered by a steam engine with a vertical boiler, located inside the front compartment, This unusual railbus was in service prior to 1910. In appearance, the body appeared to be European and contained 12 windows per side, including five doors. The motorman and passengers appeared to be ready for the run to Atchee. The small trailer was designed to carry cargo, loaded into compartments accessible through hinged doors on the sides.

LENGTH 32 FT
WIDTH 8 FT
UINTAH

206

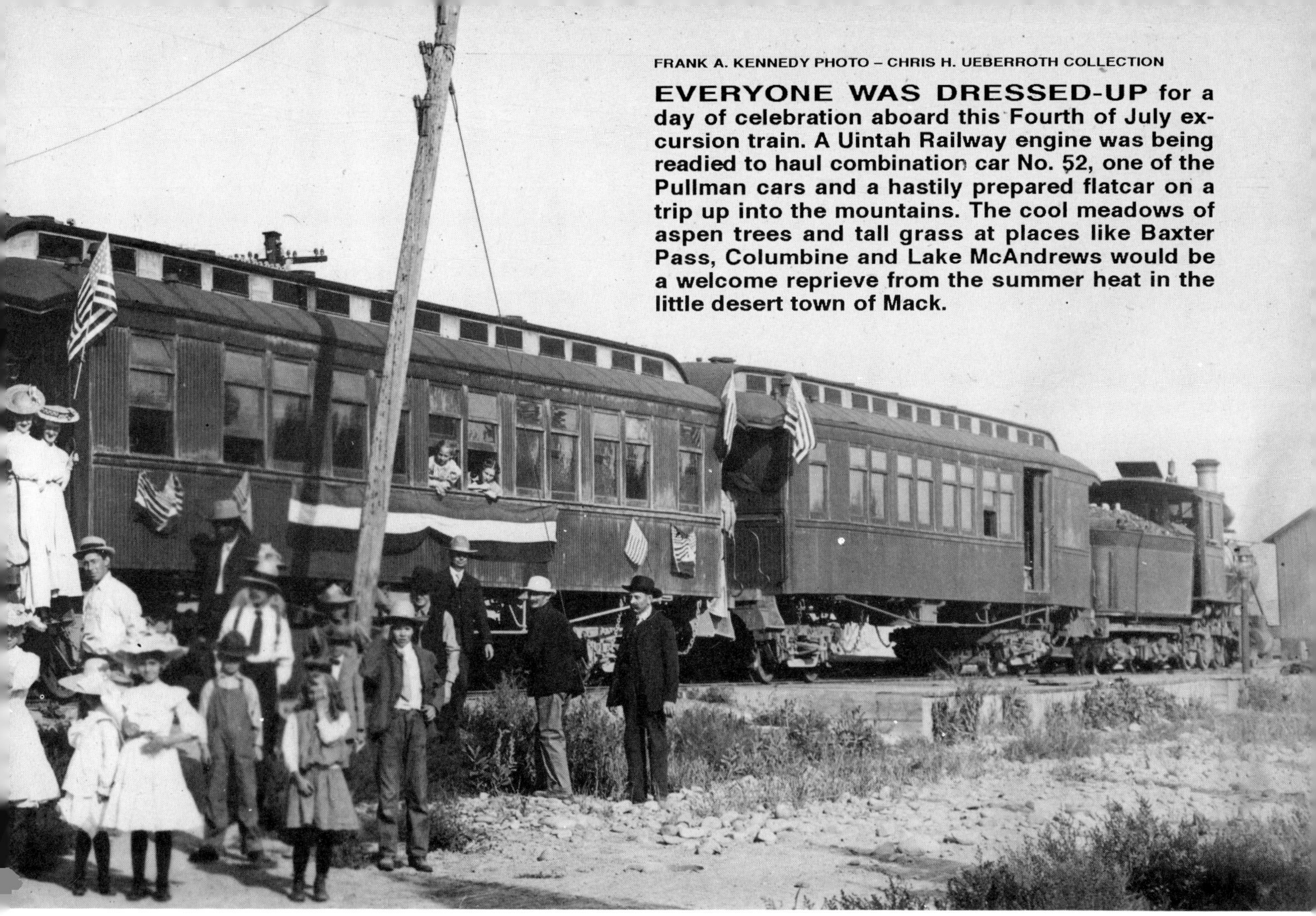

FRANK A. KENNEDY PHOTO – CHRIS H. UEBERROTH COLLECTION

EVERYONE WAS DRESSED-UP for a day of celebration aboard this Fourth of July excursion train. A Uintah Railway engine was being readied to haul combination car No. 52, one of the Pullman cars and a hastily prepared flatcar on a trip up into the mountains. The cool meadows of aspen trees and tall grass at places like Baxter Pass, Columbine and Lake McAndrews would be a welcome reprieve from the summer heat in the little desert town of Mack.

FRANK A. KENNEDY PHOTO – CHRIS H. UEBERROTH COLLECTION

THE U.S. ARMY detachment stationed at Fort Duchesne, Utah, was in the process of unloading foot lockers from Uintah combine No. 52, having reached Mack on September 15, 1911. This was the only troop train operated by the Uintah. In this view are two combines, a D&RG Pullman car and Uintah boxcar No. 206. Onlookers at the depot were looking at the spectacle, as the troopers prepared to board the standard-gauge D&RG.

CHRIS H. UEBERROTH COLLECTION

THE UINTAH MANAGEMENT took pride in their rolling stock, as evidenced here in this scene, where a car man was washing windows on combine No. 1 before departure from Mack. Mail sacks were ready to be loaded for communities up the line, as well as all of the mail for Vernal, Utah, and other towns in the Uintah Basin.

The Uintah Railway Company

THE SCENIC LINE OF WESTERN COLORADO

Trains Daily leave Mack, Colo., at 2 p. m., arrive at Dragon, Utah, 7 p.m.; leave Dragon 7:30 a. m., arrive at Mack, Colo., 12:30 noon, connecting with D. & R. G. at Mack, Colo., and with daily line of stages at Dragon, Utah, between Dragon, Vernal and Fort Duchesne with first class service.

ROUTE FREIGHT

Via DRAGON, UTAH

To insure quick service, for the following points on wagon lines:

Vernal, Utah,	Fort Duchesne, Utah	White Rocks, Utah
Theodore, Utah	Roosevelt, Utah	Myton, Utah
Moffat, Utah	Ouray, Utah	Randlett, Utah
Rangely, Colo.	Jensen, Utah	

M. W. COOLEY, General Manager, MACK, COLORADO

JOSEPH PERRY. Assistant Secretary, MACK, COLORADO

RODGER POLLEY COLLECTION

FORM 2009

THE UINTAH RAILWAY COMPANY
MACK, COLORADO

CHRIS H. UEBERROTH COLLECTION

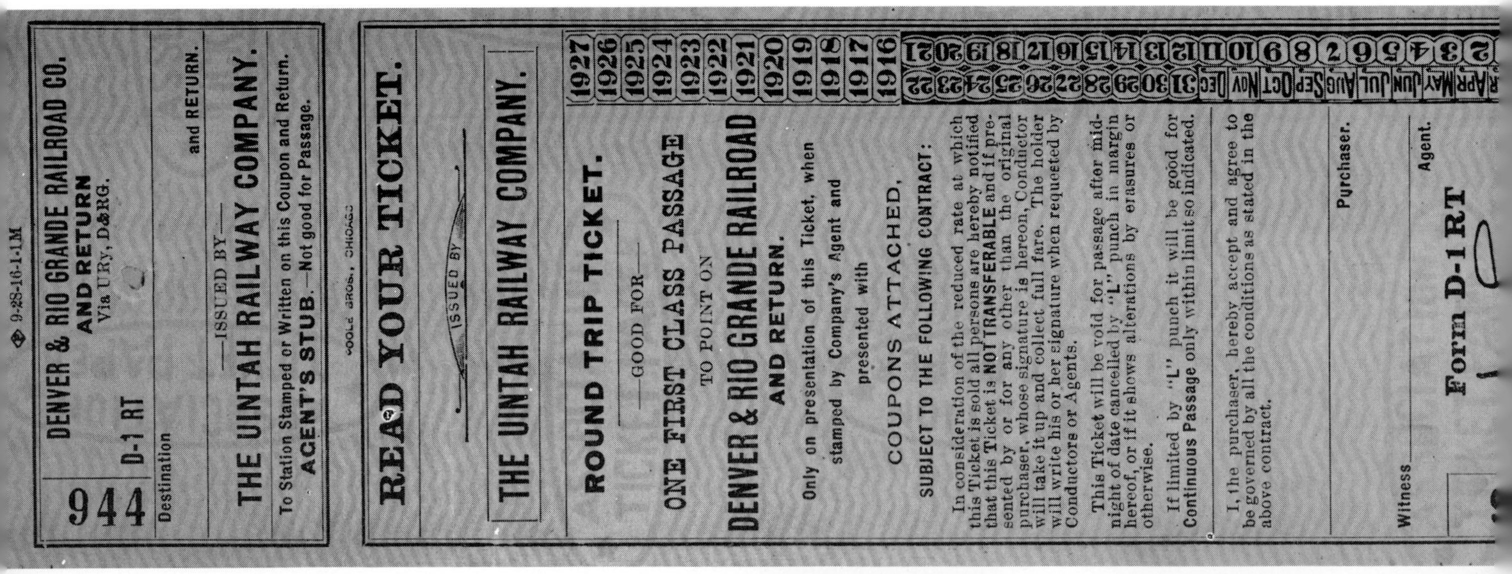

9-28-16-1-1M

944 D-1 RT

DENVER & RIO GRANDE RAILROAD CO. AND RETURN Via U Ry, D&RG.

Destination and RETURN.

—ISSUED BY— THE UINTAH RAILWAY COMPANY.

To Station Stamped or Written on this Coupon and Return. AGENT'S STUB.—Not good for Passage.

READ YOUR TICKET.

ISSUED BY THE UINTAH RAILWAY COMPANY.

ROUND TRIP TICKET.

—GOOD FOR—

ONE FIRST CLASS PASSAGE

TO POINT ON

DENVER & RIO GRANDE RAILROAD AND RETURN.

Only on presentation of this Ticket, when stamped by Company's Agent and presented with

COUPONS ATTACHED,

SUBJECT TO THE FOLLOWING CONTRACT:

In consideration of the reduced rate at which this Ticket is sold all persons are hereby notified that this Ticket is NOT TRANSFERABLE and if presented by or for any other than the original purchaser, whose signature is hereon, Conductor will take it up and collect full fare. The holder will write his or her signature when requested by Conductors or Agents.

This Ticket will be void for passage after midnight of date cancelled by "L" punch in margin hereof, or if it shows alterations by erasures or otherwise.

If limited by "L" punch it will be good for Continuous Passage only within limit so indicated.

I, the purchaser, hereby accept and agree to be governed by all the conditions as stated in the above contract.

Purchaser.

Witness Agent.

Form D-1 RT

1916 1917 1918 1919 1920 1921 1922 1923 1924 1925 1926 1927

JAN FEB MAR APR MAY JUN JUL AUG SEP OCT NOV DEC

1 2 3 4 5 6 7 8 9 10 11 12 13 14 15 16 17 18 19 20 21 22 23 24 25 26 27 28 29 30 31

DESTINATION
Via U Ry, D&RG.
On Conditions named in Contract.
ONE PASSAGE NOT GOOD IF DETACHED
944
Form D-1 RT
ISSUED BY
THE UINTAH RAILWAY COMPANY.
DESTINATION
The Denver & Rio Grande R. R. Co.
To MACK, Colo.
Via U Ry, D&RG.
On Conditions named in Contract.
ONE PASSAGE NOT GOOD IF DETACHED
944
Form D-1 RT
ISSUED BY
THE UINTAH RAILWAY COMPANY.
DESTINATION
The Denver & Rio Grande Railroad Co.
MACK, Colo. to
Via U Ry, D&RG.
On Conditions named in Contract.
ONE PASSAGE NOT GOOD IF DETACHED
944
Form D-1 RT
ISSUED BY
THE UINTAH RAILWAY COMPANY.
DESTINATION
THE UINTAH RAILWAY COMPANY.
To MACK, Colo.
Via U Ry, D&RG.
On Conditions named in Contract.
ONE PASSAGE NOT GOOD IF DETACHED
944
Form D-1 RT
ISSUED BY
THE UINTAH RAILWAY COMPANY.

ED AND NANCY BATHKE COLLECTION

A FAVORITE LOCATION for taking photographs was under the station sign at the Mack depot. These three men waited patiently for the photographer to finish his work, out in the bright afternoon sun on the southwest side of the depot. The telegraph operator in the depot received and sent messages to both the D&RGW and all points along the Uintah Railway—as far off as the Gilsonite mines at Dragon and Rainbow, Utah. Notice the train-order hoop held by the man at the right. He would hand a train order up to a trainman on a passing train, making it unnecessary for the train to stop at the station.

THREE-FOOT-GAUGE Uintah Railway tracks made their way out of the switching yard and converged into a single track. This scene was viewed in front of the Denver & Rio Grande Western's depot at Mack. The camera was facing the southeast, down the tracks, with the depot across the track opposite the long row of poplar trees that stood in front of the Uintah Hotel and office building, on the left. The picture was taken during the late 1930's, after Coxs' Mack Mercantile building—which stood at the end of the tracks—had been torn down. The backside of the Independent Lumber Company warehouse was at the far right.

RODGER POLLEY COLLECTION

MOODY PHOTO – DAN PRICE COLLECTION

THE *ROYAL GORGE ROUTE* emblem of the Denver & Rio Grande Western Railroad shows clearly on the Mack depot in this 1938 photograph. The depot had been reroofed and repainted several times during its life; notice the banded paint scheme on the chimneys, typical of the D&RGW's buff-and-brown depot livery. Window-glass breakage must have been a problem, considering the white putty present around the panes of glass. Although the Uintah Railway was about to disappear from the scene, the depot still saw its share of Denver & Rio Grande Western passenger and freight trains every day. The tall train-order semaphore was in the "stop" position, which allowed the station agent to give written orders to the standard-gauge train crews. At the same time, Uintah employees in the same room went about dispatching trains and organizing transfer shipments of Gilsonite ore from one road to the other.

TED WURM PHOTO – CHRIS H. UEBERROTH COLLECTION

THE UINTAH'S MIXED TRAIN had been spotted on the end-of-track at Mack on July 16, 1939. It was a hot day, as evidenced by the open window. The lack of money for maintenance allowed the depot to age—in need of paint and platform replacement. The red semaphore arm was in the lower (horizontal) stop position, which meant that no train could pass

the depot's train-order signal without obtaining orders. This manually controlled semaphore would be dropped to the lower (proceed) position after orders were delivered. What looks like a 1937 Packard motor car was parked near the depot, at the public-access end of the building. Notice that there was a tall mast for a radio antenna rising above the depot roof.

RICHARD A. RONZIO COLLECTION

THE REAR DOOR of the Uintah Railway's combination car No. 50 was closed, clearly showing the length of the window that provided passengers with memorable views along their journey over the line—as the scenic wonders receded. The photograph of the rear of No. 50 was taken at Mack, just a few weeks before the last passenger train ran, on May 16, 1939. The large caboose-style lanterns were permanent additions on the rear end of the car, and these marker lamps had been electrified when No. 50 was built in 1934.

COMBINATION CAR No. 50 originally was an all-wooden car, which operated on the Uintah for many years. This car was totally rebuilt in the shops at Atchee in 1934. The interior was refurbished and the entire exterior was replaced with metal sheathing. In this form, No. 50 was used as the Uintah's main passenger coach from 1934 until the closing of the line in 1939. Three times a week, this combine would be coupled to freight cars destined for Watson, Utah. These mixed trains would leave Mack on Monday, Wednesday and Friday—returning from the other end of the line the following days. This picture shows the station agent at the depot in Mack loading parcels into the baggage/L-C-L compartment of No. 50. This car usually was the only source for small shipments for the communities of Dragon, Rainbow and Watson. The curved end sill allowed for a better turning radius on Baxter Pass.

DAN PRICE COLLECTION

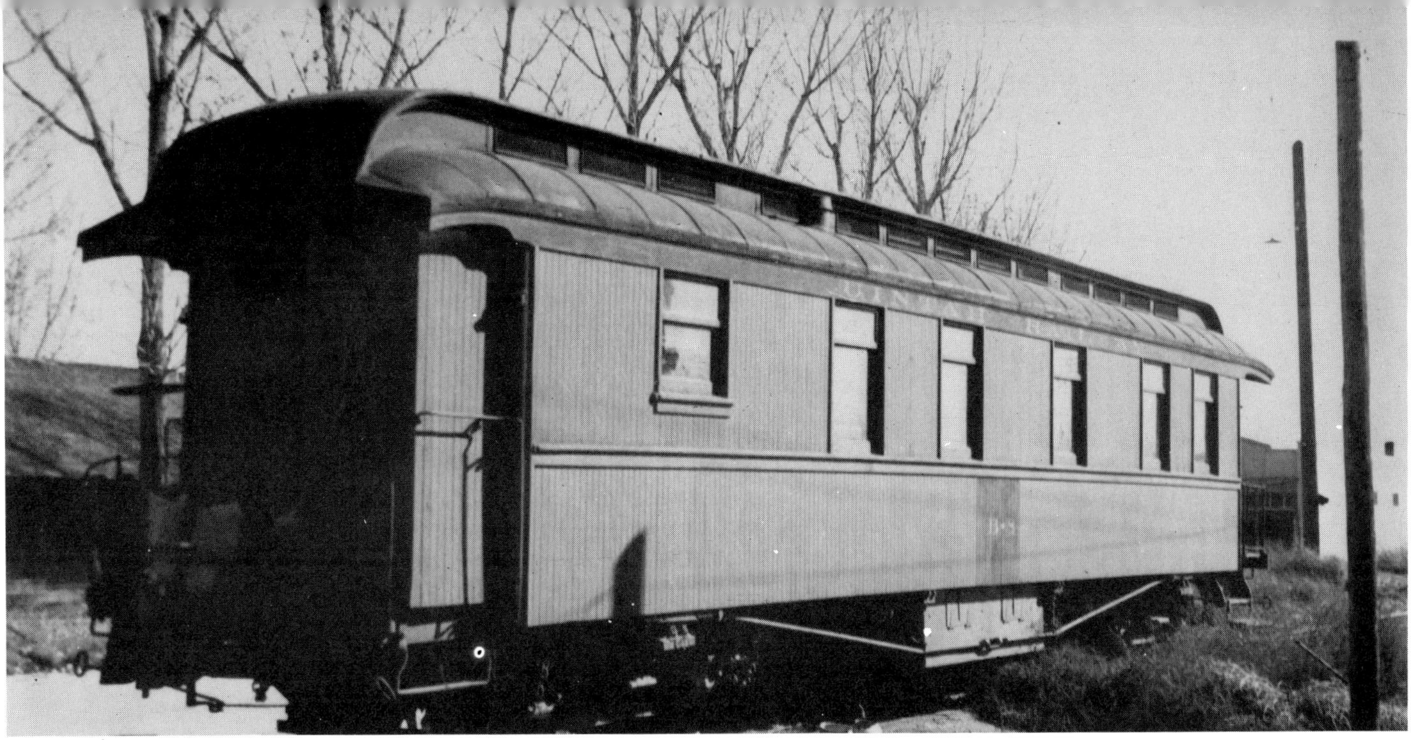

ROY F. BLACKBURN PHOTO—CHRIS UEBERROTH COLLECTION

PASSENGER CAR No. B-8 had been setout on the tracks east of the depot, in front of the Mack Hotel on November 7, 1938. This car originally was built for the Denver & Rio Grande Railway by Jackson & Sharp in 1871 or 1872, and lettered "K." This D&RG car was purchased by the Uintah Railway in 1927, and it kept the D&RGW's designation as No. B-8. This car was not used in regular passenger service; however, it was used by railway management for guests and officials on countless special occasions.

THE OBSERVATION END of the Uintah Railway's Business Car B-8 had been built with an extended platform and featured large picture windows. This allowed visiting dignitaries to enjoy the passing scenery while traveling along the line. This view was taken on November 7, 1932.

ROY F. BLACKBURN PHOTO
JOHN W. MAXWELL COLLECTION – COURTESY DAN ABBOTT

G. L. BOOTH PHOTO
RODGER POLLEY COLLECTION

A STRAIGHT-ON side view of business car No. B-8 shows the window arrangement and observation platform, as well as the food cooler (ice box) on the front platform, near the car's galley.

RODGER POLLEY COLLECTION

THIS PHOTOGRAPH of Mack was taken during the 1920's, and it makes a good picture for comparison with the first photograph portrayed in this volume, shown on Pages 24–25. The wooden water tank was still in use, and the icehouse (with a vent on the roof) was at the right. Several houses had been built on this side of the D&RGW, and trees have matured, obscuring the Uintah Hotel and office building. The combination D&RGW/URY depot was basking in late-afternoon sunlight. The Independent Lumber Company's warehouse had gone up on this side of the mainline, across the track from the Mack Mercantile building.

Form 2021—10-25—2m Mail this report to Superintendent at end of each trip.

THE UINTAH RAILWAY COMPANY
MACK, COLORADO

Conductor's Daily Trip Report

Train No.__________ Engine No.__________ Date__________________________192___

From ______________________________ To ______________________________

Called to Leave____________________ at ______________ M.. Left at__________ M.

Arrived at________________________ at ______________ M., Relieved at________M.

____

Conductor ______________________________ Brakeman ______________________________

Engineman ______________________________ Fireman ______________________________

Delays ____

RICHARD A. RONZIO COLLECTION

AN ENLARGEMENT of a D&RGW/URY Mack depot picture shows the windows and a door positioned on the telegraph-bay side of the building. The left end was used as the freight room, while a waiting room for passengers was in the right end. The D&RGW's outhouse was positioned on this side of the mainline. A modest house stood in the foreground, where one of the Uintah officials tried farming as a "side line."

RODGER POLLEY COLLECTION

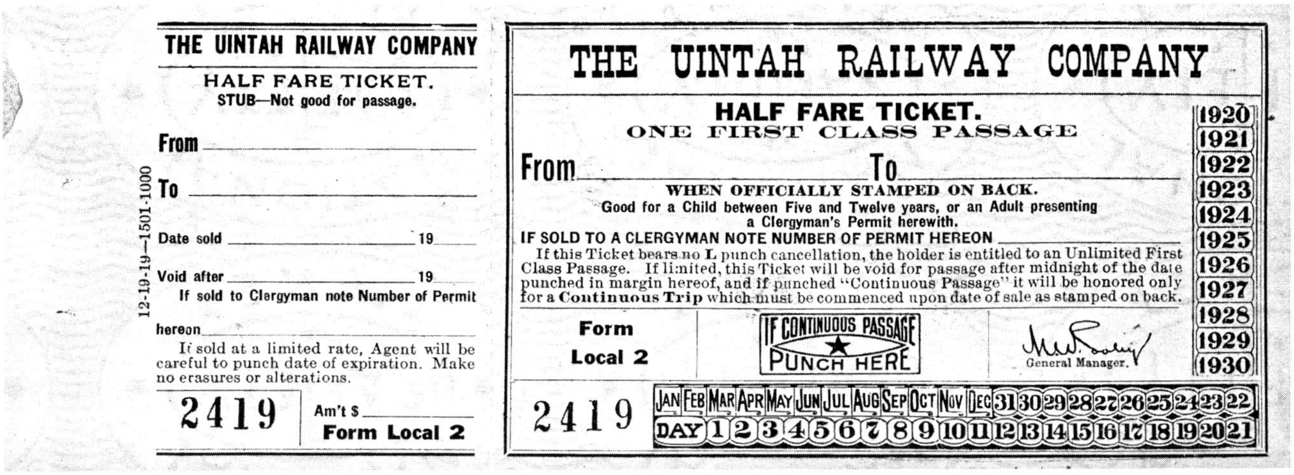

THE UINTAH RAILWAY COMPANY

HALF FARE TICKET.
STUB—Not good for passage.

From ______

To ______

Date sold ______ 19___

Void after ______ 19___

If sold to Clergyman note Number of Permit hereon ______

If sold at a limited rate, Agent will be careful to punch date of expiration. Make no erasures or alterations.

12-19-19—1501-1000

2419 Am't $ ______ Form Local 2

THE UINTAH RAILWAY COMPANY

HALF FARE TICKET.
ONE FIRST CLASS PASSAGE

From ______ To ______

WHEN OFFICIALLY STAMPED ON BACK.
Good for a Child between Five and Twelve years, or an Adult presenting a Clergyman's Permit herewith.

IF SOLD TO A CLERGYMAN NOTE NUMBER OF PERMIT HEREON ______

If this Ticket bears no L punch cancellation, the holder is entitled to an Unlimited First Class Passage. If limited, this Ticket will be void for passage after midnight of the date punched in margin hereof, and if punched "Continuous Passage" it will be honored only for a Continuous Trip which must be commenced upon date of sale as stamped on back.

Form Local 2

IF CONTINUOUS PASSAGE ★ PUNCH HERE

General Manager.

1920 1921 1922 1923 1924 1925 1926 1927 1928 1929 1930

2419

JAN FEB MAR APR MAY JUN JUL AUG SEP OCT NOV DEC 31 30 29 28 27 26 25 24 23 22

DAY 1 2 3 4 5 6 7 8 9 10 11 12 13 14 15 16 17 18 19 20 21

RICHARD A. RONZIO COLLECTION

FRANK A. KENNEDY PHOTO – RICHARD A. RONZIO COLLECTION

IN THIS SCENE, the Uintah Railway had placed function above all else for this 2-8-0. This view was photographed a few years after the view of No. 11 illustrated on Page 32, following extensive changes that had been made to the engine. The engine's cab had been replaced, and a new wooden pilot graced the front end. The engine also had a replacement tender, and she had been given a new paint job and lettering. The diamond smokestack was replaced with a capped stack, and the ornate headlight had been replaced with an electric carbon-arc type. The first No. 11 served the Uintah well for the first seven years; however, she was scrapped in 1911, when the Uintah got a brand new Consolidation from the Baldwin Locomotive Works, also numbered "11."

RICHARD A. RONZIO COLLECTION

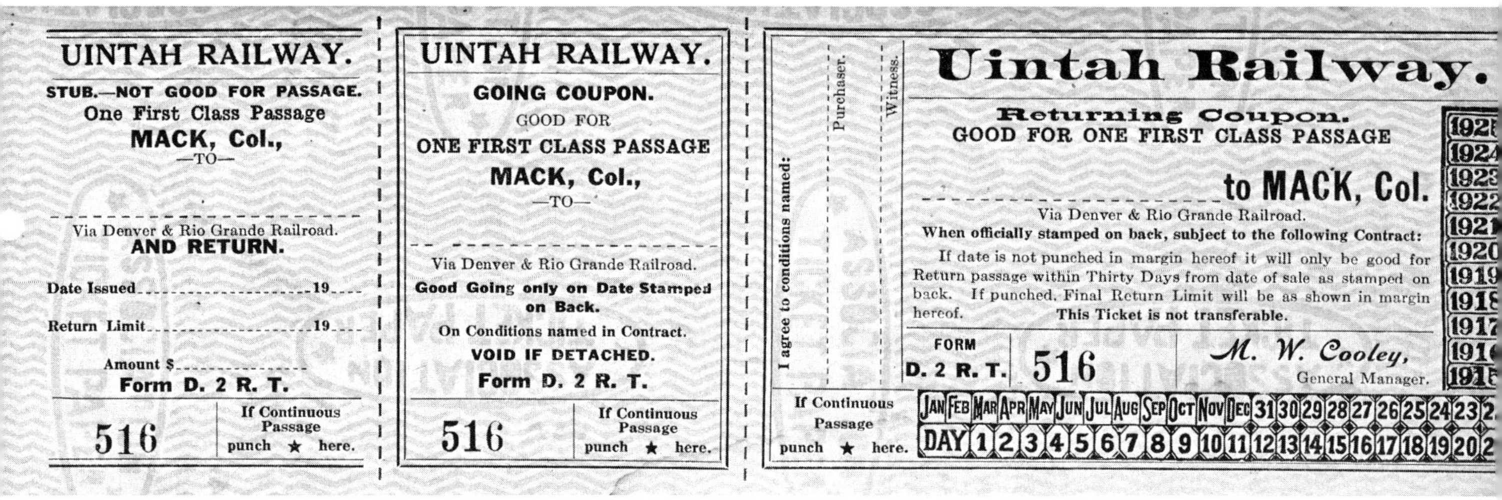

UINTAH RAILWAY.
STUB.—NOT GOOD FOR PASSAGE.
One First Class Passage
MACK, Col.,
—TO—
Via Denver & Rio Grande Railroad.
AND RETURN.
Date Issued .... 19....
Return Limit .... 19....
Amount $..................
Form D. 2 R. T.
516
If Continuous Passage punch ★ here.

UINTAH RAILWAY.
GOING COUPON.
GOOD FOR
ONE FIRST CLASS PASSAGE
MACK, Col.,
—TO—
Via Denver & Rio Grande Railroad.
Good Going only on Date Stamped on Back.
On Conditions named in Contract.
VOID IF DETACHED.
Form D. 2 R. T.
516
If Continuous Passage punch ★ here.

I agree to conditions named:
Purchaser.
Witness.

Uintah Railway.
Returning Coupon.
GOOD FOR ONE FIRST CLASS PASSAGE
.... to MACK, Col.
Via Denver & Rio Grande Railroad.
When officially stamped on back, subject to the following Contract:
If date is not punched in margin hereof it will only be good for Return passage within Thirty Days from date of sale as stamped on back. If punched, Final Return Limit will be as shown in margin hereof. This Ticket is not transferable.
FORM D. 2 R. T. 516
M. W. Cooley, General Manager.
1925 1924 1923 1922 1921 1920 1919 1918 1917 1916 1915
If Continuous Passage punch ★ here.
JAN FEB MAR APR MAY JUN JUL AUG SEP OCT NOV DEC 31 30 29 28 27 26 25 24 23 2
DAY 1 2 3 4 5 6 7 8 9 10 11 12 13 14 15 16 17 18 19 20 2

JACK POLLOCK COLLECTION

ENGINE 10 was a common sight around the Mack yard for the first 20 years. This photograph of the right side of No. 10 shows the back of the tender and one of the earlier styles of lettering the Uintah used. She was heading upgrade, toward Atchee.

FRANK A. KENNEDY PHOTO – RICHARD A. RONZIO COLLECTION

A SINGLE COMBINATION CAR usually made up the train for the Uintah Railway's passenger runs from Mack, Colorado, to Dragon and Watson, Utah. Engine 10 was used by the Uintah during the early years as its main passenger engine on the less steep portion of the line, from Mack to Atchee. These two photographs of No. 10 were taken in the yard at Mack, near the end-of-track, ca. 1917. Fireman Vic Earp was visible in the cab window, while Engineer George Lyman was standing beside the drive wheels, holding his oil can. They were waiting for a photograph to be taken of their engine, as the train's conductor was standing next to the combine, ready for departure. A Baldwin Locomotive Works builder's plate, No. 24271, shows distinctly on the side of one of the Uintah's cleanest and most highly polished engines. She had been extensively rebuilt in the machine shops at Atchee just a few years earlier—after a wreck at Bridge 1-B.

ON NEARLY EVERY DAY a small passenger train departed and returned at the little Mack depot. This picture of the northwest end of the depot shows Uintah Railway Engine 10 and combination car No. 1 ready to leave for Atchee during the morning of May 18, 1917. This locomotive had gone through some refurbishing after the bridge washout accident in October of 1911. At the very least, the locomotive had received a new cab, steam dome, sand dome and smokestack. No. 10 would be used in the desert between Mack to Atchee for the next 10 years until it was sold to the Eureka–Nevada Railway.

UINTAH COUNTY LIBRARY COLLECTION

The Uintah Railway Locomotive Roster

March 24, 1931

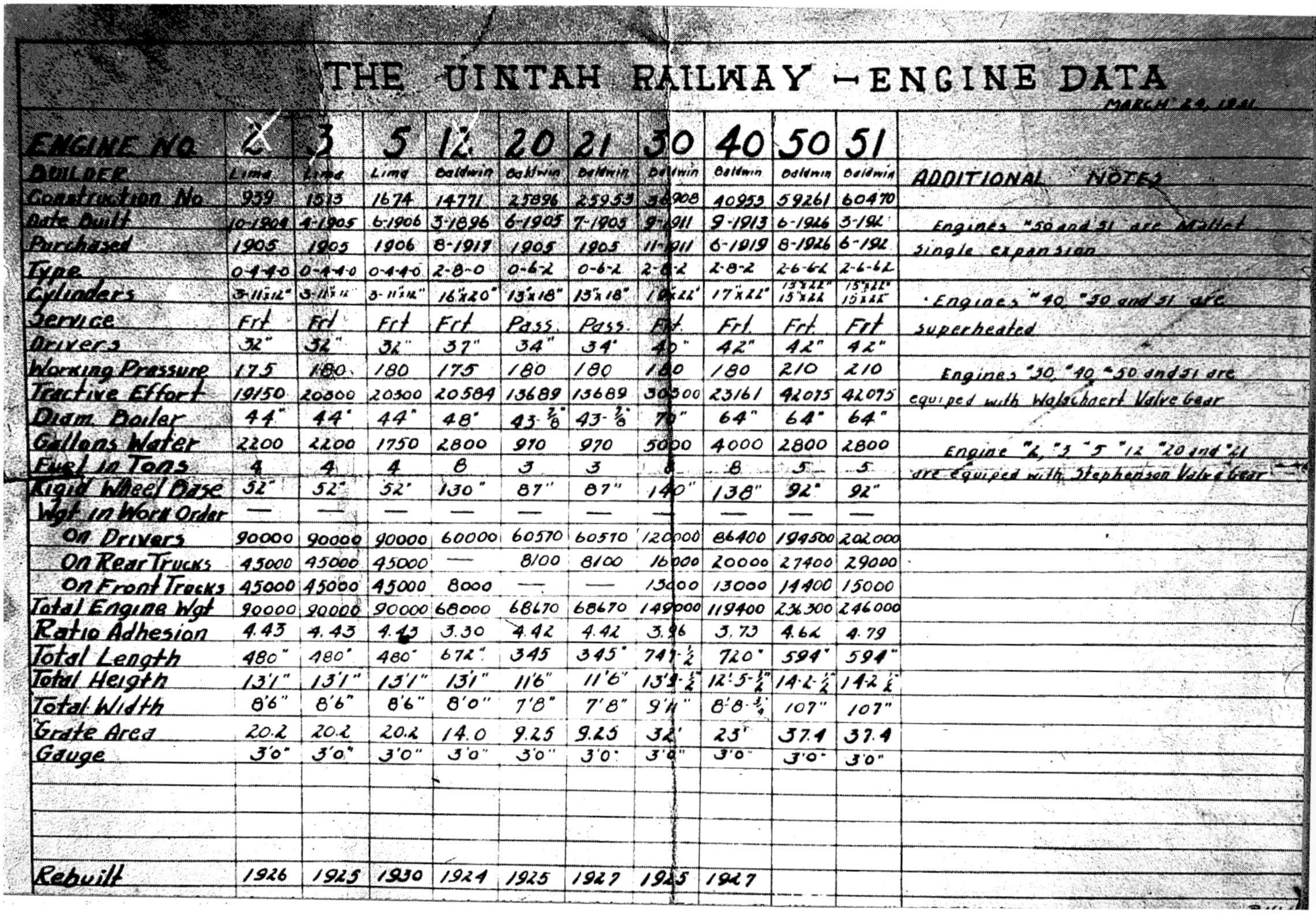

THE UINTAH RAILWAY — ENGINE DATA

MARCH 24, 1931

ENGINE NO	2	3	5	12	20	21	30	40	50	51
Builder	Lima	Lima	Lima	Baldwin	Baldwin	Baldwin	Baldwin	Baldwin	Baldwin	Baldwin
Construction No	939	1513	1674	14771	25896	25955	36908	40955	59261	60470
Date Built	10-1904	4-1905	6-1906	3-1896	6-1905	7-1905	9-1911	9-1913	6-1926	3-192
Purchased	1905	1905	1906	8-1917	1905	1905	11-1911	6-1919	8-1926	6-192
Type	0-4-4-0	0-4-4-0	0-4-4-0	2-8-0	0-6-2	0-6-2	2-[illegible]-2	2-8-2	2-6-6-2	2-6-6-2
Cylinders	3-11x12"	3-11x12	3-11x12"	16"x20"	13"x18"	13"x18"	1[illegible]x22"	17"x22"	13"x22" 15"x22	15"x22" 15"x22
Service	Frt	Frt	Frt	Frt	Pass.	Pass.	Frt.	Frt.	Frt.	Frt
Drivers	32"	32"	32"	37"	34"	34"	4[illegible]"	42"	42"	42"
Working Pressure	175	180	180	175	180	180	1[illegible]0	180	210	210
Tractive Effort	19150	20300	20300	20584	13689	13689	30500	23161	42075	42075
Diam Boiler	44"	44"	44"	48"	43-7/8"	43-7/8"	7[illegible]"	64"	64"	64"
Gallons Water	2200	2200	1750	2800	910	970	5000	4000	2800	2800
Fuel in Tons	4	4	4	8	3	3	[illegible]	8	5	5
Rigid Wheel Base	52"	52"	52"	130"	87"	87"	140"	138"	92"	92"
Wgt in Work Order	—	—	—	—	—	—	—	—	—	—
On Drivers	90000	90000	90000	60000	60570	60510	120000	86400	194500	202000
On Rear Trucks	45000	45000	45000	—	8100	8100	16000	20000	27400	29000
On Front Trucks	45000	45000	45000	8000	—	—	13000	13000	14400	15000
Total Engine Wgt	90000	90000	90000	68000	68670	68670	149000	119400	236300	246000
Ratio Adhesion	4.43	4.43	4.43	3.30	4.42	4.42	3.[illegible]6	3.73	4.62	4.79
Total Length	480"	480"	480"	672"	345	345"	74[illegible]-½	720"	594"	594"
Total Heigth	13'1"	13'1"	13'1"	13'1"	11'6"	11'6"	13'[illegible]-½	12'-5-½"	14-2-½	14-2-½
Total Width	8'6"	8'6"	8'6"	8'0"	7'8"	7'8"	9'[illegible]"	8'-8-¾	107"	107"
Grate Area	20.2	20.2	20.2	14.0	9.25	9.25	32'	23'	37.4	37.4
Gauge	3'0"	3'0"	3'0"	3'0"	3'0"	3'0"	3'0"	3'0"	3'0"	3'0"
Rebuilt	1926	1925	1930	1924	1925	1927	1925	1927		

ADDITIONAL NOTES

Engines "50 and 51 are Mallet single expansion

Engines "40 "50 and 51 are superheated

Engines "30, "40 "50 and 51 are equiped with Walschaert Valve Gear

Engine "2, "3 "5 "12 "20 and "21 are equiped with Stephenson Valve Gear

DAN ABBOTT COLLECTION

THE ENGINE DATA SHEET for the Uintah Railway dated March 24, 1931, showed 10 engines on the roster at the time. Of special interest were the columns: "Rigid Wheel Base" and "Rebuilt." The rigid wheelbase was a measurement between the center point of the front drivers and the rear drivers, creating a "rigid" line that was inflexible on curves. It was this measurement that determined the feasibility of an engine making the curves on the climb up Baxter Pass. As can be seen, Shay engines Nos. 2, 3 and 5 had small measurements, allowing them to maneuver the tight curves they were designed for in the logging and mining industries. Measurements of the two tank engines, Nos. 20 and 21, along with the big articulated Mallets also ran small enough to maneuver the curves. On the other hand, Nos. 12, 30 and 40 never would traverse Baxter Pass under their own power, because the wheels could not round the tight curves without derailing. The "Rebuilt" row at the bottom of this form shows the dates that certain engines had major service work performed at the Atchee shop. One of the reasons for buying engines Nos. 50 and 51 was the major maintenance bills that had been accumulating for the Uintah in the years before their arrival, as the Uintah's fleet of engines was getting older and in need of more and more attention.

A STRING of standard-gauge boxcars produced the backdrop for this photograph of Engine 12, as she headed toward Atchee. Rumbling out of the yard at Mack, the train included steel water car No. 020.

HELEN BANKS COLLECTION

CHRIS H. UEBERROTH COLLECTION

CONSOLIDATION No. 12 was resting on a track east of the boilerhouse, at the end of the wye. The date was 1937 for this scene, before No. 12 was sold to the Eureka–Nevada Railway that same year. She was an ex-Florence & Cripple Creek 2-8-0, the former F&CC No. 11.

UINTAH ENGINE 40 often was referred to as "Walt's baby," referring to the fact that General Manager Walter Rader loved No. 40 better than any other locomotive on the line. Even though the Uintah maintained all of its engines in great shape—compared with some other narrow-gauge carriers. Mikado No. 40 was given the lions' share of special treatment. She was polished and cleaned more often than the other engines and was re-painted more often than any others. This picture of the right side of No. 40 was taken in 1919, shortly after she was purchased by the Uintah. Originally, the tender on this 2-8-2 was built with a sloped back, but it was later changed by the Uintah to hold more water, as can be seen by the weld marks at a slant on the side of the tender. This was a good example of how shiny and well-maintained she was kept. Glossy paint, polished brass, detailed lettering and pin stripes let the engine crews and shop men know that this was the general manager's pride and joy.

FRANK DEAN PHOTO – DAN PRICE COLLECTION

RICHARD A. RONZIO COLLECTION

UINTAH'S ENGINE 40 had been spotted next to the icehouse, ready to leave Mack with a train of empty water cars and flatcars. No. 40 was highly favored by the Uintah's general manager. Notice the narrow-gauge Uintah boxcars on the transfer track, at the right of No. 40.

SCENERY UNSURPASSED

ON

The Uintah Railway

TRAINS DAILY leave Mack 8:15 a. m., arrive at Watson 12:20 p. m., connecting with automobiles on stage line, leaving Watson 1 p. m. and arriving at Vernal or Ft. Duchesne at 6:00 p. m. same day.

MODERN HOTEL AT MACK

AUTOMOBILES leave Vernal and Ft. Duchesne at 7 a. m. arriving at Watson 12 m., connecting with the Uintah Railway train leaving Watson at 1:00 p. m. and arriving at Mack at 5:00 p. m.

ROUTE FREIGHT via WATSON, UTAH

To insure quick service for the following points on wagon lines:

Vernal, Utah	Ft. Duchesne, Utah	White Rocks, Utah
Duchesne, Utah	Roosevelt, Utah	Myton, Utah
Moffat, Utah	Ouray, Utah	Randlett, Utah
Rangely, Colo.	Jensen, Utah	Independence, Utah

M. W. COOLEY, General Manager, Mack, Colorado

RODGER POLLEY COLLECTION

GERALD M. BEST PHOTO – COLORADO RAILROAD MUSEUM COLLECTION

MIKADO No. 40 was sitting in the same spot, east of the boilerhouse, waiting to find out what her fate would be. Engine 40 would not have to wait long; one year later, the engine would be at the mercy of the scrappers' cutting torch. This engine was one of only two 2-8-2's owned by the Uintah, and had eight flanged drive wheels, making it impossible for the engine to negotiate the curve at Morro Castle on Baxter Pass.

ENGINE No. 40 was sitting on the Uintah's mainline track during the late 1930's, under the eaves of the company's icehouse. The two tracks in the foreground also were Uintah tracks, used for transferring Gilsonite to the parallel tracks of the D&RGW. A small log icehouse originally located here was replaced by this structure, which was moved from Lake McAndrews. It was used to store ice for the town.

CHRIS UEBERROTH COLLECTION

40
URY
40

40

BALDWIN LOCOMOTIVE WORKS PHOTOS – H. L. BROADBELT COLLECTION

THE ENGINEER'S SIDE of the Uintah Railway Company's No. 50 is illustrated in this builder's photograph. Notice the air tanks atop the articulated engine's boiler.

THE REAR OF No. 50 shows the coal bunker of the 2-6-6-2T, with a footboard for brakemen to stand on while shunting a train. The articulated engine was delivered with a backup light, useful for wyeing or switching the locomotive at night.

THE FIREMAN'S SIDE of Uintah No. 50 shows the unusual arrangement of piping. Water for the boiler was contained in the large tanks mounted on each side of the locomotive.

BALDWIN LOCOMOTIVE WORKS PHOTO – H. L. BROADBELT COLLECTION – COURTESY CHRIS H. UEBERROTH

THE CAB INTERIOR of 2-6-6-2T No. 50 was photographed at the Baldwin Locomotive Works from the fireman's side. Firemen had a maze of valves to tend on the engine.

BILL AND SUE KNOUS COLLECTION

THERE WAS ALWAYS more to be done around the yard at Mack than just the never-ending transfer of Gilsonite from narrow-gauge to standard-gauge cars. Every piece of equipment, machinery, building material, or part that was destined for towns along the Uintah or for Vernal, Utah, had to be transferred at Mack. Every one of the Uintah's engines and every piece of equipment arrived that way, and in one way or another, departed that way. In July of 1926, and again in April of 1928, two shipments from the Baldwin Locomotive Works would forever change the way the Uintah Railway operated. The transfer gangs at Mack would have an interesting day as articulated engine No. 50 was delivered on a standard-gauge flatcar. Dual tracks on the north side of the American Gilsonite Company's warehouse provided the logical place for the large Lehigh Valley heavy-duty flatcar, No. 9952, to be spotted.

AFTER THE BIG TANK ENGINE was unsecured from the flatcar, a ramp was built, using crossties and bridge timbers. Rail then was laid to the top of the ramp and connected to the rail under No. 50. A few flatcars were then pushed to the top of the ramp, where they were coupled to the front of No. 50, and the all-important airbrakes were hooked up. After everything was checked and rechecked, No. 50 was carefully rolled down to the solid track in the yard. Flatcars were loaded with sacks of Gilsonite to keep them from buckling upward from the weight of the big engine. The narrow-gauge flatcars were utilized so the Uintah engine used to inch the heavy articulated 2-6-6-2T off the standard-gauge Lehigh Valley flatcar did not have to go up the makeshift ramp.

BILL AND SUE KNOUS COLLECTION

ALLAN C. LEWIS COLLECTION

MASTER MECHANIC Shafer, Engineer Pollock and Fireman High posed with articulated locomotive No. 50 before modifications had been applied to the 2-6-6-2T. Someone had written on the back: "Engine 50, Uintah Ry., the crookedest narrow-gauge [railway] in the world. Only two of these engines were ever built... haul 145 tons from Atchee to Baxter Pass, seven miles [upgrade], and it takes one hour to make [the run]."

Cylinders: 15x22 inches	Drivers: 42 inches
Weight on drivers:	194,500 lbs.
Weight on pilot truck:	14,400 lbs.
Weight on trailing truck:	27,400 lbs.
Total weight of locomotive:	236,300 lbs.

VARIOUS PIPES, hoses and valves were located next to the track at the end of the wye. This is where Mack's domestic water had to be unloaded from railway cars into underground cisterns, before being pumped into the steel water tank. This 1937 photograph shows Engine 50 as she was sitting at the end of the wye—after the airpumps on her head end had been relocated to both sides of the smokebox. They were attached to the head-end frame, next to the headlight. The big wooden box in the foreground contained a large hose that was placed in an engine's tender when the water supply had to be replenished.

DR. LESTER L. WILLIAMS COLLECTION

BALDWIN LOCOMOTIVE WORKS PHOTO – H. L. BROADBELT COLLECTION

LOCOMOTIVE No. 51 was sitting on the 66-degree curve at the Baldwin Locomotive Works before being shipped to Mack, Colorado, during April of 1928. The airpumps had been mounted next to the smokebox, unlike No. 50, which had them placed beneath the headlight on the pilot.

SPECIAL TRACKAGE had been laid at the Baldwin plant to test the operation of No. 51 on the same curvature as found at Morro Castle on the grade to Baxter Pass. This test also was performed on Engine 50, but with an important part of the problem missing—the 7.5-percent grade! Although by the time this picture was taken of Uintah No. 51 on the test track, all of the problems associated with No. 50's failure on the curve at Morro Castle had been rectified, it was still an in complete test without the extra binding that takes place when an engine traverses a sharp curve while ascending a steep grade. However, all-in-all, the Baldwin engineers made a breakthrough in a never-before-tried narrow-gauge technology—i.e., successfully operating an articulated engine around such sharp curves. The metal-sheathed box on top of the pilot contained extra weight (a block of concrete), permitting the pilot truck to go around the curves better.

BALDWIN LOCOMOTIVE WORKS PHOTO – H. L. BROADBELT COLLECTION

51
10160-A

BALDWIN LOCOMOTIVE WORKS PHOTO – H. L. BROADBELT COLLECTION – CHRIS H. UEBERROTH COLLECTION

ALTHOUGH THE TWO ARTICULATED locomotives were an engineering marvel and did the job they were intended to do, they were essentially one-of-a-kind prototypes that had to be modified over time for any number of reasons. The engineer's side of No. 51 is seen in this photograph at the Baldwin Locomotive Works during April of 1928. Slightly larger than her older sister, locomotive No. 50, she had a few refinements to the design, including a larger coal bunker and a better designed steam-delivery system for the front drivers, which allowed the front truck to traverse the heavy grades and sharp curves of Baxter Pass.

BALDWIN LOCOMOTIVE WORKS

HIGH-PRESSURE ARTICULATED TANK LOCOMOTIVE
2-6-6-2T TYPE

BLW CLASS 16-24 / 24-$^1/_4$-DD, 2 — **ROAD NO. 51**

BUILT FOR UINTAH RAILWAY

ENGINE

Gauge	3′0″
Cylinders (4)	15″ x 22″
Valves	Piston, 8″ diam.

BOILER

Type	Straight Top
Diameter	64″
Working Pressure	210 lbs.
Fuel	Coal

Firebox

Material	Steel
Staying	Radial
Length	$106^1/_{16}$″
Width	$66^1/_4$″
Depth, front	55″
Depth, back	43″

Tubes

Diameter	$5^3/_8$″	2″
Number	24	146
Length	18′ $0^1/_{16}$″	18′ $0^1/_{16}$″

Heating Surface

Firebox	125 sq. ft.
Tubes	1,975 sq. ft.
Total	2,100 sq. ft.
Superheater	505 sq. ft.
Grate Area	37.4 sq. ft.

DRIVING WHEELS

Diameter Outside	42″
Diameter Center	36″
Journals, Main	$7^1/_2$″ x 8″
Journals, Others	7″ x 8″

ENGINE TRUCK WHEELS

Diameter, Front	26″
Journals	5″ x 8″
Diameter, Back	26″
Journals	5″ x 10″

WHEEL BASE

Driving	22′ 10″
Rigid	7′ 8″
Total Engine	38′ 3″

WEIGHT
In Working Order

On Driving Wheels	200,400 lbs.
On Truck, Front	15,000 lbs.
On Truck, Back	20,000 lbs.
Total Engine	244,400 lbs.

Tank Capacity	2,800 U.S. Gal.
Fuel Capacity	$4^1/_2$ tons

Tractive Force	42,100 lbs.
Service	Freight

Construction No. 60,470 *Built April 1928* *Drawing No. 2*

Equipped with Type A superheater, exhaust steam-injection power reverse and air brake on all driving wheels, with two airpumps

ROBERT C. BOULTER COLLECTION

BALDWIN LOCOMOTIVE WORKS PHOTO – H. L. BROADBELT COLLECTION – CHRIS H. UEBERROTH COLLECTION

THE FIREMAN'S SIDE of Engine 51 plainly shows the original lettering style and striping.

W. C. LARSON COLLECTION

DURING APRIL OF 1928, the transfer yard in Mack again was busy with preparation, as engine No. 51 was delivered via the standard-gauge D&RGW. The engine was sitting atop a heavy-duty Lehigh Valley flatcar, which had been used a year and a half earlier to deliver sister engine, No. 50. The big 2-6-6-2T locomotive was firmly secured on the flatcar, and thick timber supports had been fashioned to shore up any instability. Care had been taken regarding all aspects of shipping the new engine. Large boards had been attached outside the windows and doors to prevent any access and insure that no glass would be broken. The backup light on the cab, as well as the smokestack, had been removed to insure a safe trip between the factory and Mack, Colorado.

THE SHOW REALLY BEGAN when the D&RGW's steam hoist, No. 026, arrived. Rated to 150 tons, the big derrick had no problem lifting the front of the large narrow-gauge engine into the air so that rail could be placed under the 2-6-6-2T (on the flatcar deck). Cable was placed around the front part of the boiler and was attached to the large hook on the derrick. A group of Uintah employees, led by Master Mechanic Scott Shafer, readied themselves with cribbing to be placed under the pilot.

W. C. LARSON COLLECTION

W. C. LARSON COLLECTION

YOU CAN LOOK underneath Engine 51's drivers as she was suspended in midair by the D&RGW derrick in the back and the cribbing in the front. Sitting on top of a stack of railroad ties was a small push car marked "B&B" (which stood for the Bridge & Building Department). The transfer crew stood next to a pile of ties and rail, ready for the call to insert the ties under No. 51. Engine No. 12 rested at the right, in front of the American Asphalt Association's warehouse.

W. C. LARSON COLLECTION

ONE MORE TIME, the front of locomotive No. 51 was lifted, to allow the cribbing to be removed. Then, the front drivers were off the deck far enough for the rails to be laid under them. At the end of the gondola, one can just make out the front set of trucks waiting to be installed under the front deck.

THIS WAS the last time the Uintah Railway received a new locomotive. In April of 1928, the largest narrow-gauge engine to be used in North America was delivered to the Uintah Railway. E. V. Earp, the engineer in the cab window, was waiting for the photographer to finish, as No. 51 simmered on the track in front of the office building.

DAN PRICE COLLECTION

W. C. LARSON COLLECTION

NOW THAT No. 51 was sitting solidly on the rails atop the flatcar, there was only one thing left for the derrick to do. Using the light cable on the end of the boom, No. 026 lifted the smokestack into place for the 2-6-6-2T, as men prepared the engine for rolling onto a makeshift trestle and then on down to the ground the same way that No. 50 was unloaded. It is interesting to note the large amount of extra equipment that followed the D&RGW's derrick to the job location. The flatcar directly behind the crane was called the "boom car," and it carried extra wheels, an entire freight truck, rail, cable, chain, frogs and every size of clevis, along with other equipment.

RODGER POLLEY COLLECTION

THE STATE HIGHWAY to Utah through Mack was temporarily blocked by Uintah articulated 2-6-6-2T No. 50. The big engine had a water car in tow, while sitting on the tail of the wye at Mack. This picture was taken as an unknown young woman was standing in the cab, pretending to be the engineer.

THE TAIL OF THE WYE always was an excellent place to get a good photograph of the Uintah engines going about their work. This was true in 1935, when the famed railroad photographer Otto C. Perry shot this image of Engine 50, simmering in front of a water car that was being drained. The locomotive was nine years old by this time, and the engine's "new shine" had long since disappeared. Many modifications had been made to the engine over time, and of interest in this view, was the line, or hose, that seemed to have been draped over the front of the smokebox.

OTTO C. PERRY PHOTO – RICHARD H. KINDIG COLLECTION

W. C. LARSON COLLECTION

UINTAH RAILCAR No. 52 was put together in 1934 as an inexpensive means for the then superintendent E. V. Earp to travel along the line. This photograph was taken in front of the Uintah Railway's office building in Mack. Railcar No. 52 had its original canvas top when this picture was shot. It was designed around a Ford Model "T" chassis and was provided brake rigging by the industrious men in the machine shop at Atchee. The credits for its design and construction are lost to time, being reported variously as "being built by" just about every one of the employees who were working for the Uintah Railway during the 1930's. The front wheels were solid and the back wheels had eight spokes each.

RAILCAR NO. 52 often was stored on the end of the wye, parked next to the boilerhouse. This picture shows the east side of the tar-papered boilerhouse, as it looked during 1938. One of the large legs of the water tank was in the background, holding up thousands of gallons of water and steel above it. Late in 1938, a single-stall enginehouse was built in this same spot, to shelter the locomotive that often was left there for days on end.

COLORADO RAILROAD MUSEUM COLLECTION

DENNIS O'CONNER PHOTO

UINTAH RAILCAR 52 was on the wye in Mack on June 29, 1939. Notice the brake rigging under the motor compartment, which "put the squeeze" on all six wheels at once. There was no foot brake in this car. Also notice the covering over the side of the cab, which nearly obscurred the motorman's view during cold or wet weather.

RAILCAR No. 52 was photographed again on June 29, 1939. The most drastic change made to No. 52 had been the replacement of all of the wheels and the addition of protective fenders. The wheels in front now had three large holes in them, while the wheels in the back had 12 spokes each—and the canvas top had been upgraded with side panels. The front of this railcar also had been changed in appearance, with the addition of a larger radiator and a spacer section between it and the hood. This utilitarian car was stationed in Mack for use by the Uintah's officials, and it was stored at several different locations in the little railway yard.

RICHARD H. KINDIG PHOTO

JOHN W. MAXWELL PHOTO – DAN ABBOTT COLLECTION

IN THIS SCENE AT MACK, Railcar 52 had been parked on the Uintah mainline, near the depot, next to a raised platform, constructed to unload wagons or trucks. Cars in the distance were in the vicinity of the transfer tracks. At some point, during the late teens or early twenties, a large expansion of the yard at Mack was built just west of the wye. These tracks became a necessity for storage of some of the Uintah's fleet of freight cars as the roster swelled in the teens—leaving certain numbers of cars unused during portions of the year. By the mid-twenties, enough flatcars were in use to supply the mines primarily with flatcars, leaving many boxcars and stock cars idle as non-Gilsonite traffic slowly dwindled, with the ever-increasing use of motor trucks to move freight into Vernal, Utah. The railway's long spur going up to the Dragon mine at Dragon, Utah, was said to have been used as a storage track for stock cars; however, it would have been an unlikely place to store empty boxcars and flatcars. The answer was the addition to the yard in this location. It is not known how many tracks were added, or how they were accessed, but changes probably were common, as witnessed by the old crossties still in the ground in the middle of this photograph.

THE RADIATOR of Uintah Railcar 52 had been clogged with insects, and the paint on the hood was starting to peel. However, on June 29, 1939, the railcar still had the look of a well-designed and maintained machine. The triangle bracket under the motorman's door was a pivot point for the hand-brake lever that went up into the cab of the railcar.

RICHARD H. KINDIG PHOTO

ROBERT HANFT COLLECTION

THE YARD at Mack normally received only a few inches of snow, while the mountains to the north of Mack may have been pounded with several feet of the white stuff. This view of boxcar No. 200 and flatcar No. 124 shows the yard on a tranquil winter day. The photographer stood on the D&RGW's mainline to shoot this view, looking toward the desert, past a small house that was on the edge of town.

COLORADO RAILROAD MUSEUM COLLECTION

THIS WAS THE NORMAL-SIZE LOAD of Gilsonite ore on one of the flatcars of the Uintah. Flatcar No. 150 was sitting in the yard at Mack under a load of approximately 180 bags of ore. This photograph was taken as the camera lens was pointed northwest at the flatcar, along with Uintah boxcars that were sitting next to the icehouse, on November 6, 1938.

THIS PICTURE shows the ends of a Uintah flatcar and boxcar, illustrating two of the unique features of the Uintah's equipment. The large round "buffer" plates were standard issue on all Uintah's cars. They helped to keep slack out of the couplers on the sharp curves and steep grades of Baxter Pass. Also, the Uintah used an above-the-deck design for their airhose system.

COLORADO RAILROAD MUSEUM COLLECTION

U.RY
124
U.RY
200

U.RY.

ROY F. BLACKBURN PHOTO – CHRIS H. UEBERROTH COLLECTION

THIS WAS THE SCENE from the depot end of the wye on November 7, 1938, as a group of empty cars was switched in the yard at Mack. The high-mounted airhoses were a unique Uintah design, which helped keep the air hoses from being frozen during the winter months. The wye headed off at a sharp angle to the right of the photograph and returned back to the transfer track, on the other side of the icehouse. This photograph shows the northeast end of the icehouse. The large sectioned door was used to load ice blocks into the icehouse in layers from the floor up. The ice was harvested at Lake McAndrews, on the other side of Baxter Pass, and shipped to Mack in boxcars. As the ice was placed in the building, a large amount of sawdust was placed around each block to ensure enough insulation for the hot summer ahead. The icehouse was a favorite place for the children in town to congregate in the summer months, as the icehouse was opened up and employees from the railroad took pieces of ice to the houses and the hotel. In December, the children were aware of another important use for the icehouse, other than storing ice. A group of employees would make a special run to the top of Baxter Pass, to cut a large number of Christmas trees. They were loaded aboard a flatcar and brought to Atchee, Carbonera and Mack, where they were "put on ice" in the icehouse. When the children in town saw the Christmas-tree train, they knew the holiday was just around the corner.

DAN PRICE COLLECTION

THIS SCENE ILLUSTRATES the difference in size of the Uintah's narrow-gauge cars and the large standard-gauge boxcars sitting on an adjacent D&RGW track. The picture was taken as the photographer aimed his camera southeast, along the Uintah's transfer track. The west side of the company's icehouse shows that it had the same style of sectioned doors on both ends.

… PHOTO – CHRIS H. UEBERROTH COLLECTION

A CLOSE-UP VIEW of the end of boxcar No. 200. This photograph was taken in Mack on November 7, 1938. As with all of the Uintah's equipment, it was well-maintained and was full of small details like the special airhose arrangement, steel buffers and a small sliding door that allowed ventilation if required. This boxcar now is on display at the Grand Junction's Museum of Western Colorado, at Fourth Street and Ute Avenue.

CHRIS H. UEBERROTH COLLECTION

UINTAH BOXCAR No. 203 and caboose No. 3 were both in good shape, sitting in front of the depot on this summer day, June 14, 1938. This boxcar shows the metal "bumpers" (buffers) and the high brake hoses, a Uintah trademark on all their revenue cars. Caboose No. 3 had its cupola redesigned at some point in time from the original straight sides to the more angled sides that can be seen here.

CHRIS H. UEBERROTH COLLECTION

TWO PICTURES of Uintah boxcar No. 212 are presented here. The upper view was taken in Mack in November 1938. Notice that the car had extra horizontal supports on the side doors. The lower photograph was taken in 1939, at the time of abandonment. Notice the missing truss rods.

RODGER POLLEY COLLECTION

UINTAH BOXCAR No. 203 was resting in the Mack yard when this side view was photographed. You can clearly see the diamond arch-bar-type trucks and truss rods on this boxcar.

RODGER POLLEY COLLECTION

GONDOLA No. 316 was sitting behind Boxcar 213 on the rip track. For some reason, this car had two shorter side supports. Since this gondola was built by the Uintah at their Atchee shops, these shorter side ribs were original, but they may have been cut shorter after repair work.

RODGER POLLEY COLLECTION

A CLOSE-UP VIEW of gondola No. 300 shows the levers on the end that were used to raise and lower the "drop bottoms" attached to the rod on the side of the car. When they were released, the hinged doors of the gondola would open to the outside of the car, letting the load empty clear of the rails.

GEORGE ROSELAND COLLECTION

ALTHOUGH THERE WERE several well-known and well-used self-powered "speeders" and railcars that traveled the Uintah's rails, there were also several of them that were not as successful. This small railcar was pictured on the section of the wye that crossed the county road. The "motorman" for this four-wheel car was located on the left-hand side of the vehicle, while the passengers were seated in the rear. This little railcar had a somewhat mysterious power plant on the right side. The large round shape on the right in all likelihood housed the power chain. There was also evidence of a removable crank by the circle marks around the center shaft that is sticking out. A lighting system and the neat appearance of the windshield, passenger seat and front guard all show that care that was taken in building the car. Several different levers are visible above the motor, and by the length of some of them, you may wonder if the passengers were not a vital part of operating the machine.

THE UINTAH RAILWAY had seven of these steel water-tank cars that were used to haul mountain spring water to the desert towns on both sides of Baxter Pass. There were two different styles built by the Uintah, but they were very similar in design. This view shows both the older No. 023 and the newer No. 018. The older styles were slightly larger and were constructed using rivets to hold the steel work together. The older style also used the older archbar-design trucks, while the newer style cars utilized the much more modern (for narrow gauge) Bettendorf freight trucks. The newer style also utilized welding instead of the more labor-intensive riveting work. These two cars had been spotted on the east side of the wye, on the curve near the depot turnout. This was where the Uintah unloaded water into a cistern for pumping up into the steel water tank.

DAN PRICE COLLECTION

KEN FRICK PHOTO

RED CABOOSE No. 3 and boxcar No. 203 sat in the mid-day sun across the road from the Uintah Railway's hotel on June 14, 1938. Caboose No. 3 was used exclusively between Mack and Atchee, because it was considered to be too heavy for the grade over Baxter Pass. It was the second eight-wheeled caboose operated by this narrow-gauge road.

CABOOSE No. 3 was used by the Uintah Railway from Mack to Atchee. This photograph of the caboose, taken in Mack, shows it to be in good shape. During later years, the cupola was rebuilt with angled side walls and grab irons on the roof. This view of the side of Caboose 3 shows the placement of the stove smokejack next to the cupola and some of the underbody detail.

RODGER POLLEY COLLECTION

THE UINTAH RAILWAY was built to the same kind of economical standards as most of the other narrow-gauge railroads in the United States. Instead of a complex enginehouse and turntable, the Uintah had a small wye to turn its trains around. The tail of the wye ran northeast from the Uintah's transfer track, and it became a natural division for the town of Mack. This picture was taken from the center of the wye during the late 1930's, looking toward engine No. 40—dwarfed by the Uintah's large steel water tank.

FRANK DEAN PHOTOGRAPH – GEORGE ROSELAND COLLECTION

"BARBER ASPHALT" was the name of the parent company of the Uintah Railway. The building adjacent to the tank housed the stationary boiler and electric generators. The track to the left of the enginehouse was used to spot a gondola of coal just high enough to empty into the coal chute for the boiler. A hostler was on duty at night to keep both the fires in Engine 40 and a boiler for the generator from going out. The boilerhouse was run during the day and created enough electricity to supply the town. At night, when the boiler was shut down, so was the electricity. Running from right to left, the county road crossed the tracks on the other side of the small bridge and trestle.

THE RAILWAY TRESTLE and road bridge were built to span the irrigation drainage ditch. This small trestle was the scene of an accident in the 1930's. A man walking home at night crossed the small railway trestle and fell through between the ties, breaking his leg. He was stranded there, unable to free himself from this position. After he yelled loudly, a group of people from surrounding residences came to the rescue, and he was promptly taken for medical attention in Fruita. The road bridge to the right of the railway trestle, allowed cars to cross the ditch and circle around to the front of the Uintah office building and Uintah Hotel. This road is still in the same location, and it is the only reminder at Mack that a railroad had the right-of-way before cars.

W. C. LARSON COLLECTION

THE STEEL WATER TANK'S base and boilerhouse were photographed from the southeast during the late 1930's. This illustration shows the backside of Engine 40, as the Mikado sat next to the boilerhouse, waiting for her next run. Gondola No. 302 was perched beside the coal bin, next to the boilerhouse. Coal was brought in from Carbonera and dumped here for use in the boiler plant. Employees of the Uintah had the right to use this coal, and local teenage boys were given the job of wheelbarrowing loads of coal to the company houses for heating and cooking.

RODGER POLLEY COLLECTION

A NICE FALL DAY during the late 1930's made it a good time to take a family picture. Although the names of the family members are not known, you can only imagine the amount of noise, steam and smoke that made up everyday life in a house that was just a few feet from the tracks. In the background, the large steel water tank towered above an engine, and several water cars were emptying their contents into the cistern, which would be pumped up into the tank.

UINTAH RAILWAY TIMETABLE

Issued in 1931

CHRIS H. UEBERROTH COLLECTION

Form 2121–8–31–200

THE UINTAH RAILWAY COMPANY

Employees' Time Table No. 25

To take effect Sunday, August 23, 1931, at 12:01 a. m.

Mountain Standard Time

Superseding Time Table No. 24 and Supplements Thereto.

For the government and information of employees only, and not intended for the use of the public.

The company reserves the right to vary from this Time Table at Pleasure.

W. L. RADER, General Manager. E. V. EARP, Superintendent.

WESTBOUND

Telephone Calls	Telegraph Calls	No. 19 Second Class Leave Daily Except Sunday	No. 1 First Class Leave Triweekly Mon., Wed. and Fri.	Miles From Mack	STATIONS AND SIDINGS
--	A		11.30 a	0	MACK D 4.2
			F 11:40	4.2	CLARKTON 7.3
			F 12:01 p	11.5	SPRAGUE 6.0
			12:15	17.5	COOLEY 2.6
----			F 12:30	20.1	CARBONERA 8.2
---		6.01 a	S 1:15 1:45	28.3	ATCHEE 1.8
		6.20	1:55	30.1	MORO CASTLE 2.1
		6.40	2:15	32.2	SHALE 2.0
---		7.20	S 2:40	34.2	BAXTER PASS 1.9
		7.35	2:50	36.1	DEER RUN 1.3
		7.45	F 3:00	37.4	COLUMBINE 2.5
---		8.05	F 3:15	39.9	McANDREWS 1.0
		8.20 a	S 3:20 3:40	40.9	WENDELLA 1.2
			3:42	42.1	SEWALL 4.5
			3:55	46.6	EAST VAC 1.4
			3:57	48.0	URADO 5.3
----	DN		4:20 4:40	53.3	DRAGON D 1.3
			4:42	54.6	COUNTRY BOY 1.1
			4:47	55.7	RECTOR 2.9
			4:50	58.6	UTE 3.5
			5:05	62.1	RAINBOW JUNCTION 0.7
—	WN		5:30 p	62.8	WATSON D
		Arrive Daily Except Sunday	Arrive Triweekly Mon., Wed. and Fri.		

EASTBOUND

STATIONS AND SIDINGS	Miles From Watson	No. 2 First Class Arrive Triweekly Tues., Thur. and Sat.	No. 20 Second Class Arrive Daily Except Sunday	Car Capacity Sidings, Locat[ion] of Water, Fuel [and] Turning Stati[ons]
MACK D 1.2	62.8	3:15 p		45 Y W
CLARKTON 7.3	58.6	F 2:20		7
SPRAGUE 6.0	51.3	F 2:05		20
COOLEY 2.6	45.3	1:50		10
CARBONERA 8.2	42.7	F 1:45		24
ATCHEE 1.8	34.5	S 1:25 12:55	11.01 a	74 Y W
MORO CASTLE 2.1	32.7	12:30	10.45	
SHALE 2.0	30.6	12:20	10.25	10
BAXTER PASS 1.9	28.6	S 12:05 p	10.01	17 Y
DEER RUN 1.3	26.7	11:45	9.30	6
COLUMBINE 2.5	25.4	F 11:35	9.15	8 W
McANDREWS 1.0	22.9	F 11:12	8.45	9
WENDELLA 1.2	21.9	S 11:05	8.30 a	22 Y W
SEWALL 4.5	20.7	10:45		7
EAST VAC 1.4	16.2	10:25		9
URADO 5.3	14.8	10:15		2
DRAGON D 1.3	9.5	10:05 9:45		36 Y W
COUNTRY BOY 1.1	8.2	9:37		7
RECTOR 2.9	7.1	9.35		7
UTE 3.5	4.2	9:25		6
RAINBOW JUNCTION 0.7	0.7	9:17		Y W
WATSON D	0	9:15 a		29
		Leave Triweekly Tues., Thur. and Sat.	Leave Daily Except Sunday	

Special Rules and Instructions

No. 1—Eastbound trains are superior to Westbound trains of the same class except:

1-A—No. 19 is superior to No. 20.

1-B—Second class and extra trains must clear the time of No. 1 and No. 2 ten minutes.

No. 2—No train will leave Mack, Dragon or Watson without clearance card.

2-A—No. 19 may leave Atchee without clearance.

2-B—No. 20 may leave Wendella without clearance.

No. 3—Train Register Books are located at Mack, Atchee and Watson.

3-A—Bulletin Books Mack, Atchee and Watson.

No. 4—Yardlimit stations are Mack, Carbonera, Atchee, Wendella, Dragon, Rainbow Junction, Rainbow and Watson.

No. 5—Trainmen must inspect and see that all retainers are turned up before starting down grade and they must assist when necessary with hand brake to insure safety of the train.

5-A—All trains must be thoroughly inspected and air-tested by car men and trainmen jointly at Atchee before leaving.

5-B—On all trains at Baxter Pass, train and enginemen must inspect train and test air; while the air is applied trainmen must turn up retainers on each car to ascertain their working condition; examine all hand brakes and confer with engineer and know that everything is in working order and in safe condition before starting their train down the heavy grade.

5-C—Enginemen and trainmen must use every precaution against the parting of trains on heavy grades. In case of trouble, train must be stopped, a careful inspection made and defects remedied if possible and report made to Superintendent.

5-D—Trainmen must not uncouple cars on grade without first testing hand brakes and knowing that they will hold. Cars placarded EXPLOSIVES or INFLAMMABLE must not be cut off while in motion.

5-E—Trainmen must be out and located where they can give signals and apply air when passing over bad track or bridges covered by slow order or otherwise, and give proceed signal to enginemen after passing, which signal must be answered by the engineer.

5-F—One member of train crew must be on the rear car of train when both ascending and descending grades. All cars in train must be closely watched by trainmen and see that wheels are not sliding and damaged.

5-G—All enginemen should run carefully and approach bridges liable to wash slowly during violent rain storms or immediately thereafter to avoid accidents.

No. 6—All employees are hereby notified that there are coal chutes, buildings, platforms and other structures located on main track and on sidings, also structures and platforms belonging to private corporations and persons located on industrial sidings and spurs that WILL NOT CLEAR a man riding on the side or top of car or engine, and all employees must protect themselves from injury in passing such structures.

No. 7—The speed of trains should be so restricted that absolute safety will be assured and the maximum speed will ordinarily be that which is necessary to make the schedule, except in territory specified enginemen and conductors may moderately increase the speed above that required by the schedule when the necessity arises, if in their judgment it would be safe and prudent, bearing in mind that safety in operation is to be given first consideration and always giving due consideration to track and surrounding conditions.

7-A—All trains must use fifty minutes descending heavy grade between Baxter Pass and Atchee, and forty minutes between Baxter Pass and Wendella.

No. 8—The following signs indicate:

S—Regular stop.

F—Flag stop to receive and discharge passengers or freight.

D—Day only telegraph office.

W—Water.

C—Coal.

Y—Wye.

No. 9—Persons accompanying live stock or other freight will be carried on any freight train handling such live stock or freight, when holding proper transportation and when permission to accompany same is covered by contract. Passengers on freight trains should be informed that cabooses will not be pulled up to platform to receive or deliver passengers or baggage.

9-A—Employees holding passes will be carried on any freight train to and from points at which trains stop, when passes are stamped "Good on Freight Trains."

No. 10—Prompt report must be made to Superintendent of all accidents and unusual occurrences.

10-A—When an accident, collision of trains or any collision of trains with vehicle or pedestrian, resulting in loss of life or injury to persons in Colorado, the superior officer, the agent or employee on the ground at time of such accident shall immediately notify the Public Utilities Commission, Capitol Building, Denver, Colorado, by telegram, the details of such accident, stating the immediate location and nature of the accident and the number of persons killed or injured. Information covering such accidents must be sent by Western Union Telegraph Company wire and all agents will accept and so transmit same, assessing charges therefor against the Railroad account.

No. 11—The following are auxiliary lines: The Rainbow branch.

11-A—Trains on the Rainbow Branch will be operated under the yardlimit rule.

No. 12—Enginemen must sound whistle approaching all crossings.

EXTRACTS FROM REVISED STATUTES OF UTAH

8203. (4203.) Placing Freight Car in Rear of Train.—Every person who, in making up or running railroad trains, places or runs, or causes to be placed or run, any freight car in the rear of passenger cars is guilty of a misdemeanor, and, if loss of life or limb results from such placing or running, is guilty of felony. The term "freight car" as used in this section does not include a baggage, express or mail car.

ENGINE RATING

WESTBOUND

From Mack, Colo., to Carbonera, Colo.:

Class Eng. 12	210 tons
Class Eng. 20 and 21	125 tons
Class Eng. 40	300 tons
Class Eng. 50	525 tons
Class Eng. 51	525 tons

From Carbonera, Colo., to Atchee, Colo.:

Class Eng. 12	175 tons
Class Eng. 20 and 21	90 tons
Class Eng. 40	225 tons
Class Eng. 50	525 tons
Class Eng. 51	525 tons

From Atchee, Colo., to Baxter Pass, Colo.:

Shay Class Eng. 2, 3 and 5	60 tons
Class Eng. 20 and 21	35 tons
Class Eng. 50	90 tons
Class Eng. 51	90 tons

From Rainbow Junction, Utah, to Rainbow, Utah:

Shay Class Eng. 2, 3 and 5	105 tons
Class Eng. 30	110 tons
Class Eng. 50	200 tons
Class Eng. 51	200 tons

EASTBOUND

From Watson, Utah, to Dragon, Utah:

Class Eng. 20 and 21	250 tons
Class Eng. 30	650 tons
Class Eng. 50	1150 tons
Class Eng. 51	1150 tons

From Dragon, Utah, to Wendella, Colo.:

Class Eng. 20 and 21	100 tons
Class Eng. 30	325 tons
Class Eng. 50	525 tons
Class Eng. 51	525 tons

From Wendella, Colo., to Baxter Pass, Colo.:

Shay Class Eng. 2, 3 and 5	105 tons
Class Eng. 20 and 21	50 tons
Class Eng. 50	185 tons
Class Eng. 51	185 tons

STAFF

J. I. McClain — Assistant Secretary and Treasurer
R. H. Wilson — Engineer and Purchasing Agent
H. S. Shafer — Master Mechanic
E. C. Dunlap — Car Foreman
Dr. C. F. Montgomery — Company Physician

H. M. PHILLIPS, Chief Dispatcher

OTTO C. PERRY PHOTO – RICHARD H. KINDIG COLLECTION

MANY AVID RAILFANS and railroad photographers made their way to Mack, Colorado, during the summer of 1939 as news of the Uintah's pending abandonment spread across the country. Articulated engine No. 50 was pulling a long train of water cars, boxcars and stock cars out of the yards at Mack on May 9, 1939. The 2-6-6-2T was in prime shape as the train rolled past the stockpens located west of town, on the south side of the mainline. The small loading chute and corrals were used by the Uintah's infrequent movement of horses and sheep. At the end of the train, combination car No. 50 was making one of the few remaining passenger runs over the line to Watson, Utah, and back to Mack. One week later, the same engine and combination car would make the last scheduled passenger run over the line. This picture was taken as Otto C. Perry was looking east toward Mack, where you can see trees that had grown up around Boyden's post office and Crew's grocery in the background.

THE PHOTOGRAPHER CAPTURED this view of a stranded automobile. He was looking south toward the Uintah's mainline just northwest of Mack. This small wash was typical of the many desert drainages that the Uintah crossed on its ascent to Baxter Pass. The ground was made of decomposed shale and was easily eroded in rainstorms. This automobile had fallen victim to the erosion in one of the many washes in the area. The crumbling rock had fallen apart under the weight of the car, leaving the rear passenger side wheel spinning uselessly. The team of horses probably had been brought in to pull the early-day vehicle out of the wash. Just behind the horses, you can make out the long trestle that went up to the infrequently photographed loading chute at Mack. Unfortunately, the horses were standing in the line of sight, hiding the largest part of the structure. Although the contractors who built the Uintah Railway probably were overjoyed at the short time they took for grading the line through the desert west of Mack, the Uintah's section foremen were less enthusiastic about the time it took their roadbed to deteriorate.

RICHARD A. RONZIO COLLECTION

Chapter 2

The Desert and... CARBONERA

AT MILEPOST 20, the Uintah Railway's trackage arrived at Carbonera, Colorado, the only settlement between Mack and Atchee. Carbonera is a Spanish word that means "coal mine," and it was here that the Uintah Railway's parent company, the Gilson Asphalt Company, had its own wholly owned and operated coal mine.

Early during the development of the Grand Valley around Grand Junction and Palisade, miners and entrepreneurs had fanned out into the Bookcliff Mountains in search of coal. In the late 1800's coal was the fuel of choice to do almost anything. Individual houses and businesses relied on coal furnaces, and any kind of industry utilizing steam for power needed large quantities of coal. Throughout several different levels in the strata of sandstone and shale that made up the Bookcliff Range, there were seams of coal of varying size and quality. Small individual coal mines sprang up north of the towns of Palisade, Clifton, Grand Junction and Fruita, all vying for the sometimes lucrative business of supplying fuel.

It is unclear exactly when the coal seam at Carbonera was found and claimed, but it was almost certainly before the Uintah Railway was built. Although there was no mining done in any large quantities before the Uintah arrived, there may have been activity there in some limited scale. The long wagon haul to Loma would be the most obvious drawback, but current topographic names in the area such as "Buniger Ridge" show a link to early individual miners in the area. As preliminary surveys were done in the area, the coal lands at Carbonera were bought with the initial capital investment to build the railway.

Preparation for mining and building the incline and coal tipple started in earnest when the rails reached Carbonera in late April of 1904. By the end of May, the Uintah's first engines had been delivered and were used to push the construction over Baxter Pass to the Gilsonite mines. Coal was soon coming down the mountain side on a regular basis, supplying the many coaling stations along the right-of-way with their first batch of the black fuel. Gondolas full of coal were dispatched to Mack, Atchee, Deer Run,

(Continued on Page 113)

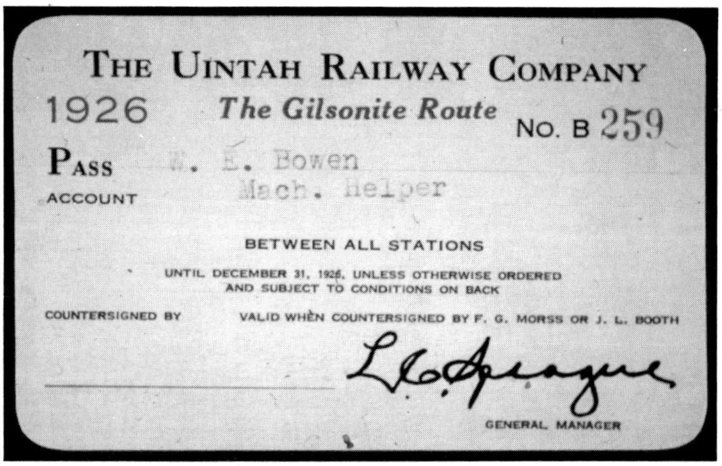

The Uintah Railway Company

1926 *The Gilsonite Route* No. B 259

Pass W. E. Bowen

Account Mach. Helper

BETWEEN ALL STATIONS

UNTIL DECEMBER 31, 1926, UNLESS OTHERWISE ORDERED AND SUBJECT TO CONDITIONS ON BACK

COUNTERSIGNED BY VALID WHEN COUNTERSIGNED BY F. G. MORSS OR J. L. BOOTH

L C Sprague

GENERAL MANAGER

UINTAH RAILWAY

Mileposts 0 to 2

— Circa 1915 —

By Rodger Polley

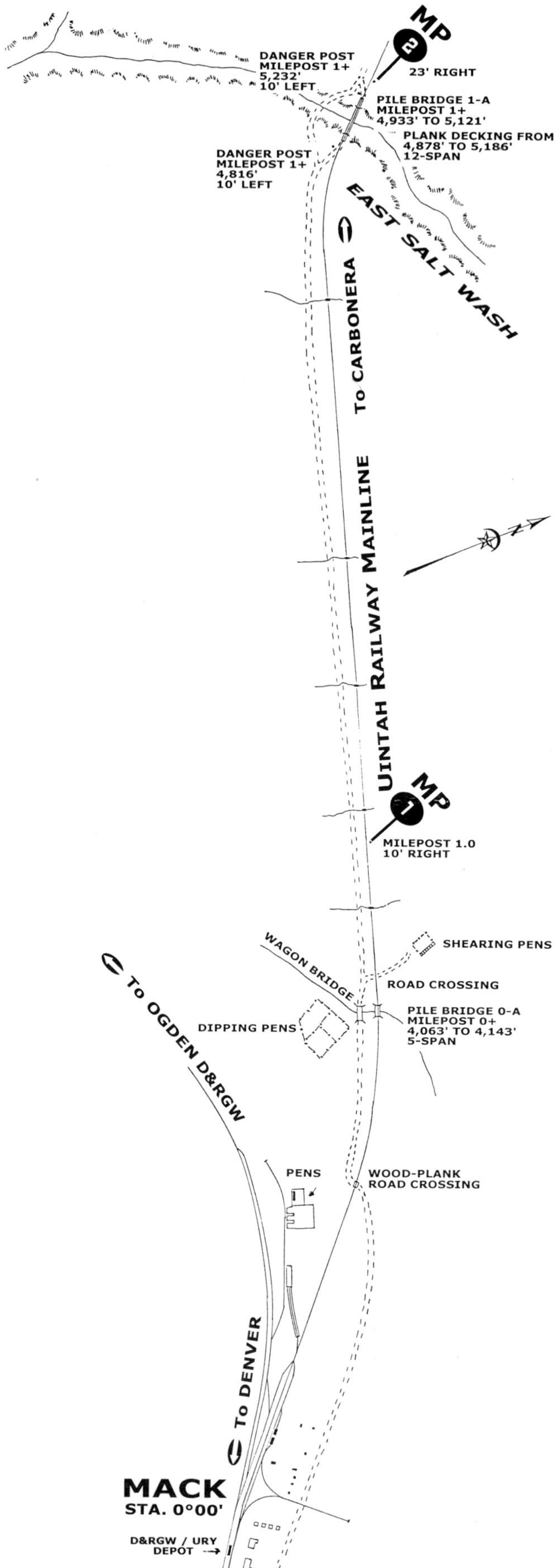

Map at Bridge 0-A...

Just west of Mack, the D&RG(W) track headed southwest away from Mack and toward the Colorado River Canyon – down a smaller gorge that was created by Salt Wash Creek. The Uintah Railway followed the old narrow-gauge route of the D&RG as it headed west, toward the desert. The wooden-pile span, Bridge 0-A, had been built across a narrow, deep wash that ran from the area north of Mack to Salt Wash to the south. This wash would eventually be used as an irrigation water overflow and usually ran water most of the year, causing it to be deeper than most of the desert washes. The dipping and shearing pens were used for sheep that were driven to this spot for shearing from all over the Bookcliff Range. The tracks were almost level, heading slightly downgrade toward East Salt Wash.

As the Uintah track went around the curve east of Bridge 1-A, the grade became sharply steeper downgrade to the trestle. From this trestle to the top of Baxter Pass, the grade continuously ascended, on an ever-increasing slope. Bridge 1-A was one of the longest trestles on the Uintah, owing to the washouts from the wreck here in 1911. The area just east and west of the bridge was a wide flat wash, and it needed a long fill on both ends to stay above the level of the ever-flooding wash. This was the only place that the Uintah crossed East Salt Wash, as it would follow the course of West Salt Wash from here to Atchee.

RICHARD A. RONZIO COLLECTION

THE PILE DRIVER had arrived at Bridge 0-A, located just northwest of Mack. The wash that ran under the bridge was located in a natural crevasse that had been enhanced by the run off from irrigation water that was brought into the area by local farmers. It never had a lot of water come through it like some of the larger washes in the area; however, water could be expected to flow here almost all year, lending itself to be a deep fairly narrow ravine. Pile Driver 001 was in front of Shay No. 1, followed by an old engine tender mounted on a short flatcar, with a load of bridge poles. The pile driver was utilizing steam from the Shay to operate a small steam engine, which provided power for the hoist system for the "hammer." With the addition of more bridges on the extension to Watson and Rainbow, as well as the need for higher maintenance for the bridges on the older part of the line, the Uintah was due a much-needed piece of work equipment. This was made evident with the deaths of two employees in the wreck at Bridge 1-A during the fall of 1911. The Uintah's Pile Driver 001 was built by the machine shop at Atchee late in 1911 to help with the ever-increasing need to rebuild and reinforce the original wooden trestles.

Wendella and Dragon to fill the coaling facilities that were built at these locations for refueling the locomotives. This initial busy activity would soon level off to a more predictable and monotonous schedule of production. Winter months were always a little bit busier than the summer, as individual heating needs in the many coal furnaces and cook stoves along the right-of-way increased demand. For the most part, though, production levels were tied in to how many operating locomotives and stationary boilers the Uintah Railway and Gilsonite mines had in operation.

The "town" itself was probably the most isolated and inhospitable location on the railroad's right-of-way. Although the small Gilsonite camps at Country Boy and Rector, Utah, were similar in non-existent comforts such as running water and electricity, Carbonera was the farthest away from any other town on the railroad that had a store, school, post office, or any other meager necessity. Although one could get mail and groceries from the conductor on a passing passenger train, running water and electricity would never be seen in Carbonera, Colorado. A telegraph was

(Continued on Page 120)

RICHARD A. RONZIO COLLECTION

AFTER THE BODIES were removed, onlookers got this view of the wreck at Bridge 1-A on the afternoon of October 1, 1911. Boxcar No. 202 was hanging over the edge of the trestle, perched on its brake cylinder and bent truss rods. Under the pile of timbers, Boxcar 202's trucks can be seen in the wreckage. Notice that the boxcar still has the stencil marks from the American Car & Foundry Company of St. Louis, Missouri, just to the left of the doorway. Engine 10, with three of its drivers showing, can be seen near the edge of the water. This picture was taken as the photographer looked northwest, facing the muddy plain left by the flood in East Salt Wash Creek. The Bridge & Building Department gang's caboose had been spotted at the end of the bridge while repairs were being made at the wreck site.

THE WRECK AT BRIDGE 1-A

ALL ALONG the Uintah Railway on September 30, 1911, an unusual amount of rain came down in the Bookcliff Range. The scheduled passenger train from Dragon to Mack was due in that evening at 5:00 p.m., but due to the heavy rain along the right-of-way, the track was in rough shape. The mixed train, consisting of a boxcar full of ice and the passenger car, changed locomotives at Atchee and headed out into the dark, stormy night toward Mack. Ahead of them at the section house at Salt Wash, a group of section hands had headed down the track toward Mack, looking for sections of track that could be dangerous, which could use some quick repairs. Water and mud were especially hard on the section of track from Carbonera to Mack, owing to the loose soil and lack of vegetation in the dry desert region.

Somewhere south of the Salt Wash siding, C. M. Kipros stayed behind to finish up working on the piece of track, and he let the rest of the crew continue on to Mack. A few hours passed before the headlight of No. 10 came into sight, as she moved ever so cautiously down the track toward the waiting section hand. As the train crept forward, Mr. Kipros hopped onto the front pilot of the engine to get a ride into Mack and to help identify any pieces of bad track should they come across any. Ahead of the train, the section crew on a handcar crossed several bridges and

WATER WAS RISING AGAIN in this picture, as salvage operations continued. All manner of block and tackle has been spread across the wash to salvage what the company could from the wreck. One of the lines has been hooked to the axle of boxcar No. 202's truck, which was lying next to the engine's tender. A group of onlookers, including some of the officers of the railway from Mack, had gathered on the east embankment to watch the proceedings and weigh important decisions as to what should be done next. In the foreground, the destructive erosion from the wash can be seen firsthand. What was solid ground just days before, had been turned into a loose piece of undercut soil, ready to fall into the wash.

RICHARD A. RONZIO COLLECTION

UINTAH RAILWAY
U.RY.
202
202

UINTAH RAILWAY

RICHARD A. RONZIO COLLECTION

THE BRIDGE & BUILDING CREW'S work train, complete with tool cars and flatcars full of materials, had moved down from Atchee and was spotted on the northwest end of the bridge. The work gang on the bridge was fastening cable from the boxcar to the caboose, preparing to drag the derailed car up onto the trestle, where it would be pulled off to Atchee for repairs. The set of trucks from boxcar No. 202 that had fallen in the wash had been removed from the side of the engine's tender and dragged across the wash in the mud, and they can be seen between the pilings just over the top of the men standing on the tender.

watched as the normal trickle in West and East Salt Wash creeks swelled to the size of small rivers. They passed over Bridge 1-A an hour ahead of the following passenger trains, and without a hint of any problems to the structure of the bridge, they headed on in to the warmth and comfort of the settlement at Mack.

ENGINEER GEORGE LYMAN and fireman Charles Pasano were taking it slow and careful as they headed into Mack, approaching the trestle over East Salt Wash at Bridge 1-A about 1:00 a.m. In the hour that had passed since the section crew crossed the bridge, the current in the wash had shifted to undermine the bridge abutments on the southeast end of the bridge. Unfortunately, the crew's timing could not have been worse. If they had crossed an hour earlier, the bridge would still have been structurally sound. An hour later and the bridge might have been altered enough to visibly see the damage from the track level. The engine crew strained to see the trestle ahead of them, as engineer Lyman cracked the throttle of No. 10 open enough to gently pull the load over the roaring waters beneath them.

The southeast end of the bridge was not visibly altered from the storm, and as far as the crew knew, there was no danger as they crossed over the halfway point of the wash and headed for solid ground only a few feet ahead.

AS ENGINE No. 10 rolled out onto the final two spans of the bridge, the pilings gave way and the engine immediately toppled into the creek below. The earth-work that held the final set of pilings in place had been washed away. The coupler between the locomotive and the rest of the train gave way, allowing the boxcar and passen-

RICHARD A. RONZIO COLLECTION

BEFORE THE BRIDGE could be rebuilt, the Uintah's Engine 10 and her tender had to be moved out of the mud. The locomotive was rolled from her left side to an upright position and was moved upstream several feet. Then, the tender was rolled onto its top—minus its trucks—and it was rolled over again on its right side. Both of these pieces of equipment were in this position on Friday, October 6, when another large storm caused the creek to dam up behind the engine and tender, washing out the bridge and an additional part of the embankment for the second time. This photograph was taken after the second washout, and the now much longer bridge had been rebuilt. The 2-8-0 had been completely covered in the resulting mud and sediment, and the work crews labored in the muck to free the equipment. The engine was in sad shape, having her cab crushed and her smokestack, sand dome, headlight and bell torn off in the wreck, or removed by the wrecking crew. The tender's superstructure had been separated from the frame during the wreck. The flow of East Salt Wash shifted to the west side of the newly cut wash, leaving a large pile of sediment in a mound in the middle of the wash, which could be seen for years.

ger car to escape the same fate. Mr. Kipros, who had been riding on the front pilot, and fireman Pasano, riding on the left side of the cab, were instantly crushed by the weight of the engine as it pushed down into the water and mud below.

PASSENGER KATHERINE TAWNEY reported the following to the *Grand Junction Daily Sentinel:*

> Suddenly we felt a slight jar, the noise of the crashing bridge and the scream of the engineer. Instantly the conductor called that the engine had gone down. It was very dark and he looked out and saw merely a blacker mass in the water and the surrounding dark.

The passenger car and boxcar had remained sitting on the north end of the trestle. The front trucks of the boxcar had fallen out from underneath the car, leaving the front of the car hanging precariously over the brink. Engineer Lyman was trapped in the wreckage and was badly injured. The cab of the engine had been crushed and the engineer's side was barely sticking out of the water. Mixed in with pieces of broken bridge timbers and having the tender of the engine pushed up near the cab, engineer Lyman found himself with a broken leg and a dislocated shoulder. With his good arm, Lyman pulled himself to the top of the wreckage, and having survived the wreck, now had to endure the possibility that the current would wash him away.

> Engineer Lyman was brave and plucky, despite his injuries and the strength of the powerful current which threatened to tear him loose from his engine and sweep him down the wash. He coolly directed the Greeks [section hands] toward his rescue, and though every effort seemed to be put forth, it was nearly an hour before he was taken from the water. I am not a good judge of distances, but I remember that it took a bell cord and four ropes from trunks in the baggage car to reach him in the engine. The rescuers worked heroically, and, handicapped by the darkness

RICHARD A. RONZIO COLLECTION

THE FRONT OF ENGINE 10 shows how deeply it was embedded in the mud. The original link-and-pin coupler had been chained in an attempt to pull the engine from the mud. Pieces of vegetation and other debris had embedded themselves in every part of the locomotive. The large hose slung across the front pilot of the 2-8-0 probably was being used to try to pump out the water from the muddy pond the work crew had created—before No. 10 could be retrived.

and the current, they could make but slow progress. The engineer could not have lived at all if he had lost his hold.

AS HELP FROM MACK ARRIVED, Lyman was finally pulled out of the wash, and he was quickly transported to the hotel at Mack, where his wounds were stabilized for the trip to St. Mary's Hospital in Grand Junction. As the daylight broke, no one else knew that section hand Kipros had also lost his life in the crash. His legs could be seen in the water near the front of the engine, but it was believed that he was the fireman Charles Pasano.

During the rest of Sunday October 1, 1911, Uintah Railway workers spent their time reacting to the wreck. The waters continued to run higher than normal that morning, although not nearly at the same level as when the engine had tried to cross the wash. The first order of business was to get the 14 passengers in the passenger car transported across the wash. As the water subsided to a nearly normal level, the passengers were lead down the stream to a point that was only a few inches deep. Only a few hours earlier, the current at that point had been four feet deep! At Atchee and Mack, the Bridge & Building Department crews were assembled along with extra help from section gangs all along the right-of-way to help in righting the engine and rebuilding the bridge. The passenger car was pulled back up the line to Salt Wash siding, and a couple of B&B outfit cars and a couple of flatcars that had been filled with timbers and supplies were pushed down the line to the site of the wreck.

ON MONDAY, the work began in earnest to get what was believed to be the body of the fireman out from under the engine. By that afternoon, the body had been removed, and owing to the fact that fireman Pasano was known not to have been wearing a coat, and it was discovered that the body had one on, it became apparent that this was C. M. Kipros – and the fact that two employees had died in the wreck first became known. Pasano was still under the bulk of the engine, buried in the wash. Strictly physical labor was used to move the engine and tender out of the wash, with no help from the conveniences of modern tractors or a derrick. Pulleys and "come-alongs," in conjunction with shovels and pry bars, were used to move the heavy locomotive in the wash.

DAN PRICE COLLECTION

THE UINTAH'S RAILCAR 52 is seen here in the desert, somewhere west of Mack, with Superintendent Vic Earp behind the "wheel." The railcar was stationed at Mack and was used by Mr. Earp almost exclusively for inspections and business along the line. This photograph shows it with its canvas top up, solid front wheels and its original radiator configuration.

THE BODY OF FIREMAN PASANO was recovered on Tuesday afternoon, after the locomotive had been moved somewhat. After the body was recovered, the focus turned to the rebuilding of the bridge. The debris from the wreck had been removed, and the boxcar was lifted onto another set of trucks and pulled off the trestle. All day Wednesday and Thursday, the trestle was rebuilt toward the embankments at each end of the bridge. The emphasis was on rebuilding Bridge 1-A, and very little attention was paid to trying to get engine No. 10 out of the wash, but unfortunately, it was the failure to remove the locomotive that would prove to be the undoing of the bridge once again. On Friday, another fierce thunderstorm rolled across the Grand Valley, and it brought even more rain down the canyons of the Bookcliff Range. At Bridge 1-A, the engine and her tender created a large diversion to the current of the water as it raced down the wash.

The engine was lying perpendicular to the current, and the water quickly started eroding the east bank away from under the trestle again. During the first flood, this erosion was limited to the natural forces of the wash and the course it had taken for many years – causing the embankment to erode just enough earth to cause the wreck. Due to the new diversion in the stream, the second flood created more power and force to erode the bank. The engine and tender, acting like a dam, caused the water to erode more than 50 feet of dirt on the southeast end of the bridge. Not only did the engine create a much larger span to be crossed, but the flood waters silted in around the engine and tender, totally erasing the gains that the work crews had made during the last week. Again, the Uintah Railway was brought to a standstill; however, thankfully, there were no more injuries from the washed-out bridge. With much toil and labor, the trestle was rebuilt and Engine 10 was hauled out of the wash. The bridge was now one of the longest bridges on the line, going from 110 feet to 188 feet. The 2-8-0 was totally rebuilt at the shops at Atchee and saw 16 more years of service on the Uintah, until she was sold to the Eureka-Nevada Railway in 1927.

hooked up at the mine foreman's house, but other than that, there was not even the sound of a radio in the entire history of this small mining town. Water and ice were delivered by the railway to a cement cistern in the ground and a small dugout-style ice house south of the coal tipple.

Here, the precious substance was hauled by hand to the houses to be used sparingly to drink, wash and clean. A small, windowless bath house was constructed out behind the old wooden cabin, and here, the miners bathed and washed-up after a grit-filled day in the mine. The mine was quite dirty compared to today's washed-down environment. Since there was no water to wash down the walls and face of the mine, coal dust permeated everything inside and near the entrance to this small mine. Outhouses lined the base of the mountain behind the houses, and trash was deposited in West Salt Wash across the track from town. Frequent flash floods in the normally dry wash "hauled the trash away." With all of this, coupled with the oppressively hot summers, equally cold winters, gnats, snakes, etc., Carbonera was not considered a nice or easy place to raise a family. Very few families stayed in Carbonera, and bachelors and young men starting off were the most common laborers to be found. Several families did raise their children there at one time or another, but most times, the family would end up living in Mack or Fruita. The miners would see their families only on weekends.

Dynamite, carbide and other explosives or chemicals, which needed to be kept away from the prying eyes and mischievous hands, were kept in a small cement dugout on the hill behind the section house. A large padlock on a small iron door ensured that the explosives would not be used inappropriately. Mine foreman Bennett kept a large brood of chickens behind his house, and the Bennetts and the boarding house were seldom in need of eggs. Also behind the houses, up along the hillside, was a large six-stall garage that was built by the men of the town after most of the family men ended up getting automobiles.

Transportation in and out of Atchee was primarily by the railroad up into the 1920's. The road from Mack to Atchee passed right through Carbonera, but until automobiles were seen more often in the 1920's, horse-drawn wagons were the only way to get back down to Atchee on the unimproved road. Even as cars became more common, the trip was impossible during the winter, as the snow and mud made the road impassable. Even during the summer, frequent flash floods caused people to get stuck in the desert for hours at a time. One resident recalled that as a child, he spent the night in the desert near Salt Wash Siding, waiting for the water to subside in order to get their automobile across the muddy creek.

The actual mine was a straight-forward uncomplicated operation. The coal seam was four feet to seven feet thick and was parallel between two layers of sandstone. The seam exited the side of the mountain about 200 feet above the bottom of the valley. Two small crevasses, eroded over time, left a small "finger" of the mountain sticking out into the valley. On the north side of this hill, the least steep area for some distance was chosen to run the incline to the coal seam. At the level of the coal seam in the bottom of this narrow, steep canyon, the miners leveled off an area big enough to build the brake- and cable-house tool shed, and tracks for the mine's tram cars. The tracks came

(Continued on Page 140)

FRANK A. KENNEDY PHOTO – PAUL AND BETTY GARBER COLLECTION

THE COST OF LABOR utilized to rebuild Bridge 1-A and the slow speed required to accomplish it, showed the Uintah's management that more modern equipment was needed to fight the continual washouts and bridge rebuilding that occurred along the line. The train wreck at this trestle also shed light on the fact that many of the original bridges along the line had not been con-

structed to a standard that could hold out against the elements. Many spans along the line were in excess of 30 feet and, because of the wreck, it was decided that the standard for most of the bridges on the line would be a maximum of 16 feet between spans. Within several weeks, the work shop at Atchee had built Pile Driver 001. This view of the pile driver was photographed on Bridge 1-A, taken by Frank Kennedy as the "hammer" fell on a new timber pile. A crewman was steadying the bottom of the new pile. This Bridge & Building gang's work train was being hauled by Shay No. 1 on the same day as the pictures of the same train taken at Bridge 0-A. (See Page 109.) Notice how the older west end of Bridge 1-A had been reinforced with truss rods.

THIS WAS THE SCENE at Bridge 2-B northwest of Mack on December 13, 1923, after the one-car passenger train running toward Atchee derailed during its Thursday-morning run. The accident happened when the rear truck on the passenger car hit a piece of broken rail just east of the bridge. Here, the wreck was viewed from the track, looking west, as a waiting train was sitting on the track west of the bridge. The passenger car was on its right side, while on the far right of the picture, a push car has been pulled off the tracks onto the embankment next to the tracks. The derailed car was pulled along until the rear truck fell off the north side of the bridge, but it was dragged the length of the trestle. By the time the passenger car reached the west end of the short bridge, the rear truck and platform hit the dirt embankment, causing the car to uncouple from the engine and roll over onto its side.

JACK AND CHARLES POLLOCK COLLECTION

EIGHT PASSENGERS were badly injured in the wreck, including General Manager James Hood and Superintendent Lucius Sprague. "Major" Hood was thrown from the rear platform of the car and suffered a broken leg. He would not recover from the injuries, and later died in the hospital in Grand Junction of complications due to his diabetes on Sunday, December 16, 1923. The other seven passengers who were injured had varying injuries from bruises and lacerations, to broken ribs. The brakeman on the combination car, Frank Wagaman, received deep cuts on his head while riding in the baggage section of the combine. A child's snow sled, complete with sharp metal runners, had been stored on an upper shelf for its trip to Dragon. As the combine turned over on its side, Mr. Wagaman was hit in the head with the runners of the sled, and initially, he was thought to be the most severely injured of the passengers. However, he recovered quickly and was back in service on the train in a few weeks. Passengers had to be transported to Mack for immediate medical attention at the Uintah Hotel. The more serious cases were then taken to St. Mary's Hospital in Grand Junction and to the hospital in Fruita. Notice the difference in the length of the ties in the foreground. Owing to the fact that the Uintah's management had firm plans to eventually bridge the White River north of Watson, tunnel Baxter Pass and widen the Uintah to standard gauge, section crews south and north of Baxter Pass were supplied with standard-gauge ties for replacement along those portions of the route.

JACK AND CHARLES POLLOCK COLLECTION

BRIDGE 2-B WAS DESIGNED differently than many of the other small trestles along the line. This small wooden trestle had a single set of pilings in the center of the small wash. Large timbers bridged across the pilings to the abutments. Underneath, poles angled from the pilings back to the embankment for stability. The well-worn and shiny wheels of the front set of trucks on the combination car were in contrast to the missing rear trucks that came off in the derailment. A train consisting of one of the Consolidations and two of the old wooden water cars sat just west of the bridge. These water cars were used infrequently after the Uintah built the rectangular steel cars in 1916, but from this photograph, it is evident there were still two in existence in 1923.

THE UINTAH'S BRIDGE 2-A was located on a small local wash, which drained the shale hills to the north. The broad flat area between Bridges 1-B and 2-A was full of tamarack brush and salt grass, which grew in this often-wet drainage area. The Uintah's track was on a fill most of the way across this lowland section of the terrain to the west of Bridge 2-A, where it ran along the base of a section of shale hills north of the track. Bridge 2-B was a short bridge that crossed over a minor drainage, and it was the location of a wreck in 1923. Just west of this trestle, the tracks turned slightly westward from the northwestern course to follow the line of the shale hills to the north. Bridge 3-B was in the middle of a flat area, and it was used as a drainage to the flat land north of the right-of-way. Even though there was not much wash to be seen, a good rain on the nearby hills could leave several feet of water standing behind the grade. The road crossing the track went north to Carbonera and Atchee. The road following the Uintah's trackage worked its way to the left, across the desert to Cisco, Utah. Sheep pens and corrals, along with a small cabin and shed, were southwest of Milepost 4.0. Large tracts of open grazing lands to the west of this area were utilized by the sheep men and are still used for their livestock today. A mile or two to the southwest, a ridge of sandstone and shale was covered in cedar trees, and for the next 10 miles this ridge was visible as the western horizon.

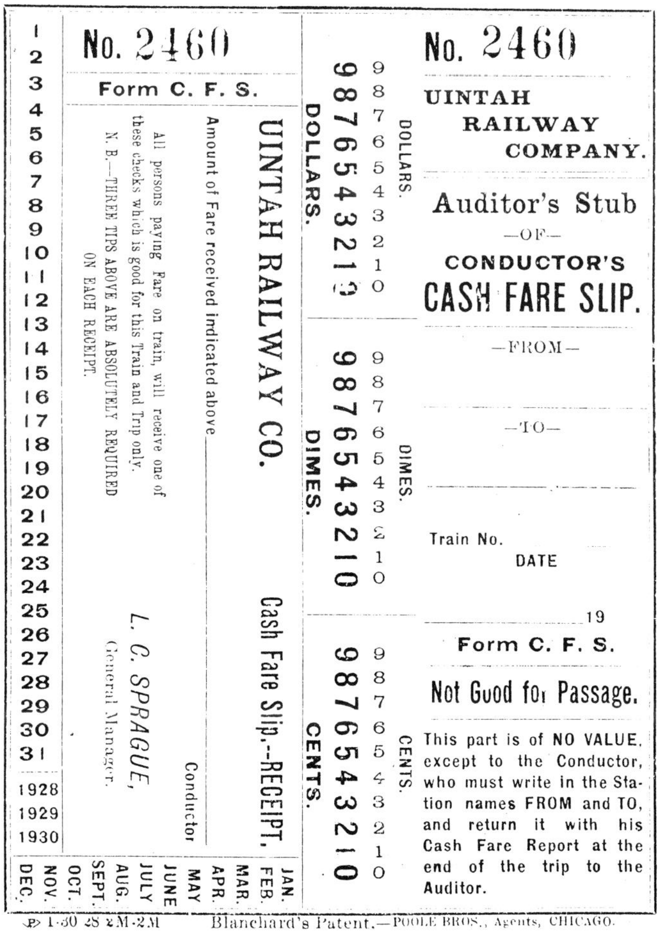

1 2 3 4 5 6 7 8 9 10 11 12 13 14 15 16 17 18 19 20 21 22 23 24 25 26 27 28 29 30 31
1928 1929 1930
JAN. FEB. MAR. APR. MAY JUNE JULY AUG. SEPT. OCT. NOV. DEC.

No. 2460
Form C. F. S.
UINTAH RAILWAY CO.
Cash Fare Slip.--RECEIPT.
Amount of Fare received indicated above
Conductor
All persons paying Fare on train, will receive one of these checks which is good for this Train and Trip only.
N. B.—THREE TIPS ABOVE ARE ABSOLUTELY REQUIRED ON EACH RECEIPT.
L. C. SPRAGUE,
General Manager.

DOLLARS. 9 8 7 6 5 4 3 2 1 0 DOLLARS.
DIMES. 9 8 7 6 5 4 3 2 1 0 DIMES.
CENTS. 9 8 7 6 5 4 3 2 1 0 CENTS.

No. 2460
UINTAH RAILWAY COMPANY.
Auditor's Stub
—OF—
CONDUCTOR'S
CASH FARE SLIP.
—FROM—
—TO—
Train No.
DATE
19
Form C. F. S.
Not Good for Passage.
This part is of NO VALUE, except to the Conductor, who must write in the Station names FROM and TO, and return it with his Cash Fare Report at the end of the trip to the Auditor.

1-30 2S 2M-2M Blanchard's Patent.—POOLE BROS., Agents, CHICAGO.

RODGER POLLEY COLLECTION

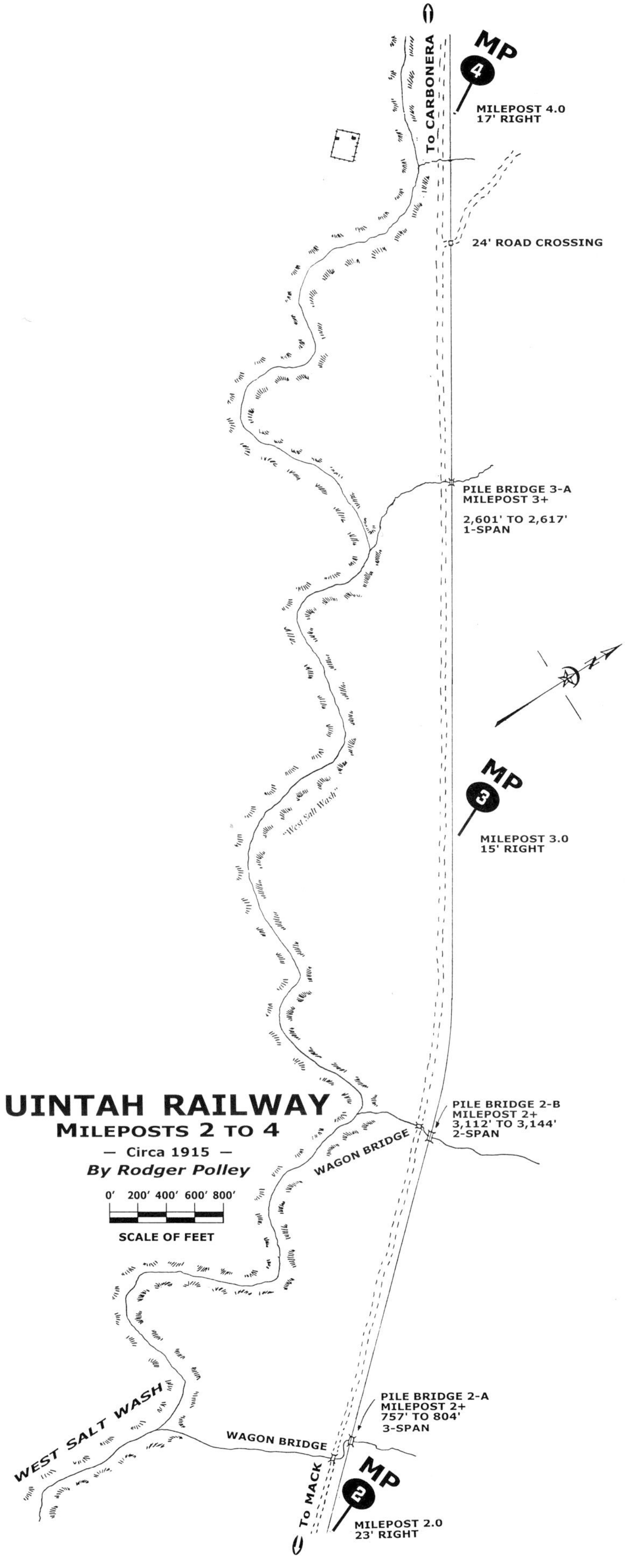

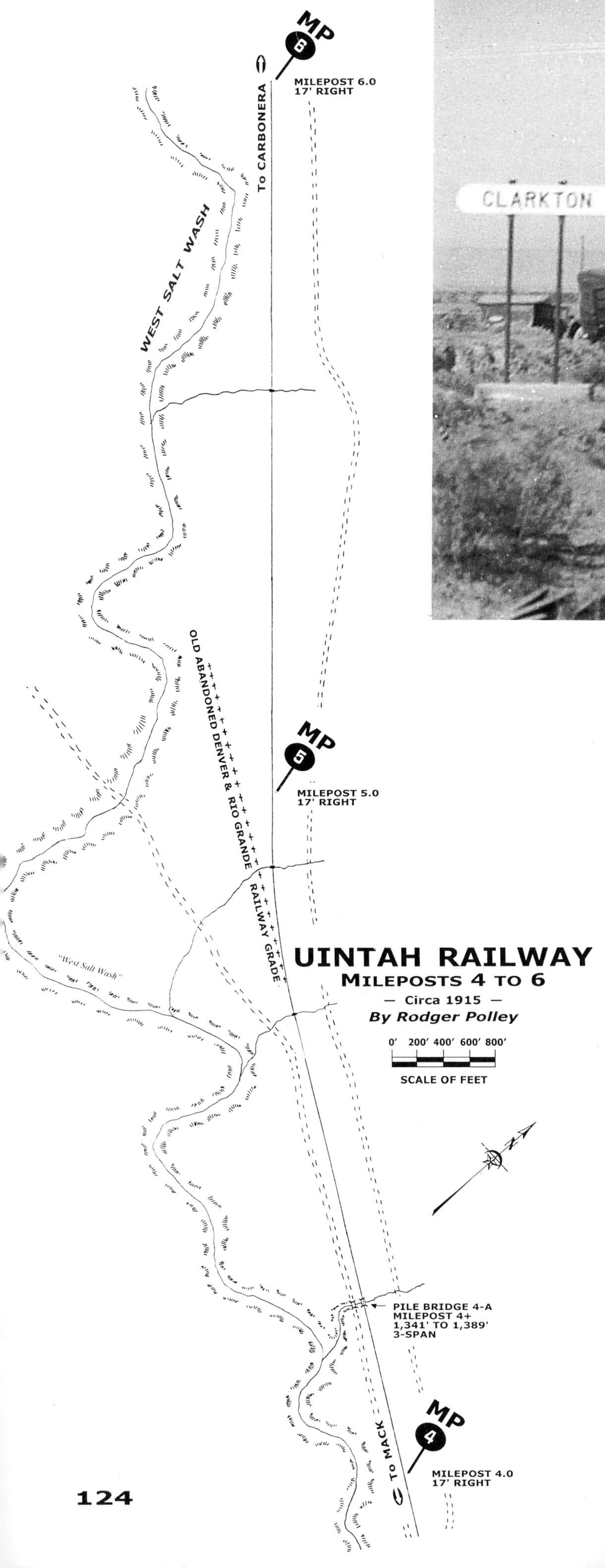

BETWEEN MILEPOST 4.0 and Bridge 4-A, a small spur was constructed during the early 1920's to provide the local farmers with a means of shipping sugar beets to the processing plants in Grand Junction and Delta. The siding was named Clarkton and was used almost exclusively for shipping sugar beets to market. Pile Bridge 4-A was constructed across a small side wash, while the dirt road to the south was traversed by a wooden bridge. About three-quarters of the way to Milepost 5.0, the Uintah left the path of the old narrow-gauge Denver & Rio Grande and turned more to the northwest, along the east bank of West Salt Wash. West of this location, the abandoned D&RG roadbed slowly snaked its way up a large swell in the topography that makes up the western horizon. With this large upheaval of shale and sandstone to the west, and the shale plateau around Mack to the east, the Uintah traveled on a tangent path for the next few miles in the desolate dry wash that made up the valley of West Salt Wash. The desert to the west of here was spotted with sagebrush and ground cover; however, on many of the hillsides, there was little vegetation whatsoever. Between Mileposts 5.0 and 6.0, the barren shale hills just east of the track later were turned into productive farms, complete with green fields and trees along irrigation ditches, thanks to the canal projects that were completed in the area during the late teens. However, West Salt Wash was the natural end of the line for the irrigation ditches, and the Uintah Railway's right-of-way made a distinct boundary between productive farmland to the east and barren desert to the west.

JIM McCOY COLLECTION

CLARKTON SIDING—about a quarter-mile below Milepost 4.0—was built in 1920 or 1921 to accommodate the farming community of New Liberty and the local farmers west of Mack. The Holly Sugar Company in Delta, Colorado, started production in 1920 and contracted with farmers all over the Western Slope of Colorado to obtain sugar beets for their plant. At quite a few locations on the D&RG's tracks serving the communities of western Colorado, the Holly Sugar Company built loading facilities to load the beets into D&RG cars. Somewhere near Mack, the D&RG had one of these loading facilities. The Uintah, in turn, would bring the sugar beets in from Clarkton to Mack and dump them at this loading site. Some confusion exists as to exactly how and where this facility was operated. Some stories tell of a third rail in between the D&RG rails to the location on the siding at Mack, where the D&RG loaded cars. It is unknown how long this spur was, or how many cars it held. This view of the Clarkton siding was taken looking upgrade northwest toward Salt Wash and Carbonera. Heavy transportation work for farmers during the 1920's usually was still done mostly by horse and wagon, although there are a few early automobiles in the photograph. The siding was north of the mainline, where a small building was built and the self-powered conveyor was brought in during harvest to load the cars.

Form 2110–1[illegible]27–1000.

THE UINTAH RAILWAY COMPANY

BRAKE AND SIGNAL EQUIPMENT

Engine No. 40

Tested and Cleaned at Atchee, Colorado Shop

Air Compressor, Orifice Test	Date 11-18-38 192
Main Reservoir, Hydrostatic Test	Date 8-18-38 192
Air Gauges Tested	Date 11-18-38 192
Distributing Valve Cleaned	Date not used 192
Control Valves Cleaned	Date 8-18-38 192
Triple Valves Cleaned	Date 8-18-38 192
Reducing Valves Cleaned	Date 8-18-38 192
Straight Air Double Check Valves Cleaned	Date not used 192
Dirt Collectors Cleaned	Date not used 192
Brake Cylinders Cleaned	Date 8-18-38 192

RODGER POLLEY COLLECTION

JIM McCOY COLLECTION

AS THE LOADING was completed, the farmers were paid for a full load! A close-up view of three cars shows that the second car back was actually a Uintah Railway flatcar, which had posts in the side pockets and large 2 x 12-foot timbers were placed on edge to produce another gondola. Innovation on the part of the railway allowed its small fleet of gondolas to be used for the important hauling of coal to points along the railway, while still being able to service the farmers' needs around Mack. This was the heyday for Clarkton siding. With the importation of sugar cane from various points around the world, the sugar-beet industry died in western Colorado, and by the late 1920's, there was little—if any—traffic from the siding at Clarkton.

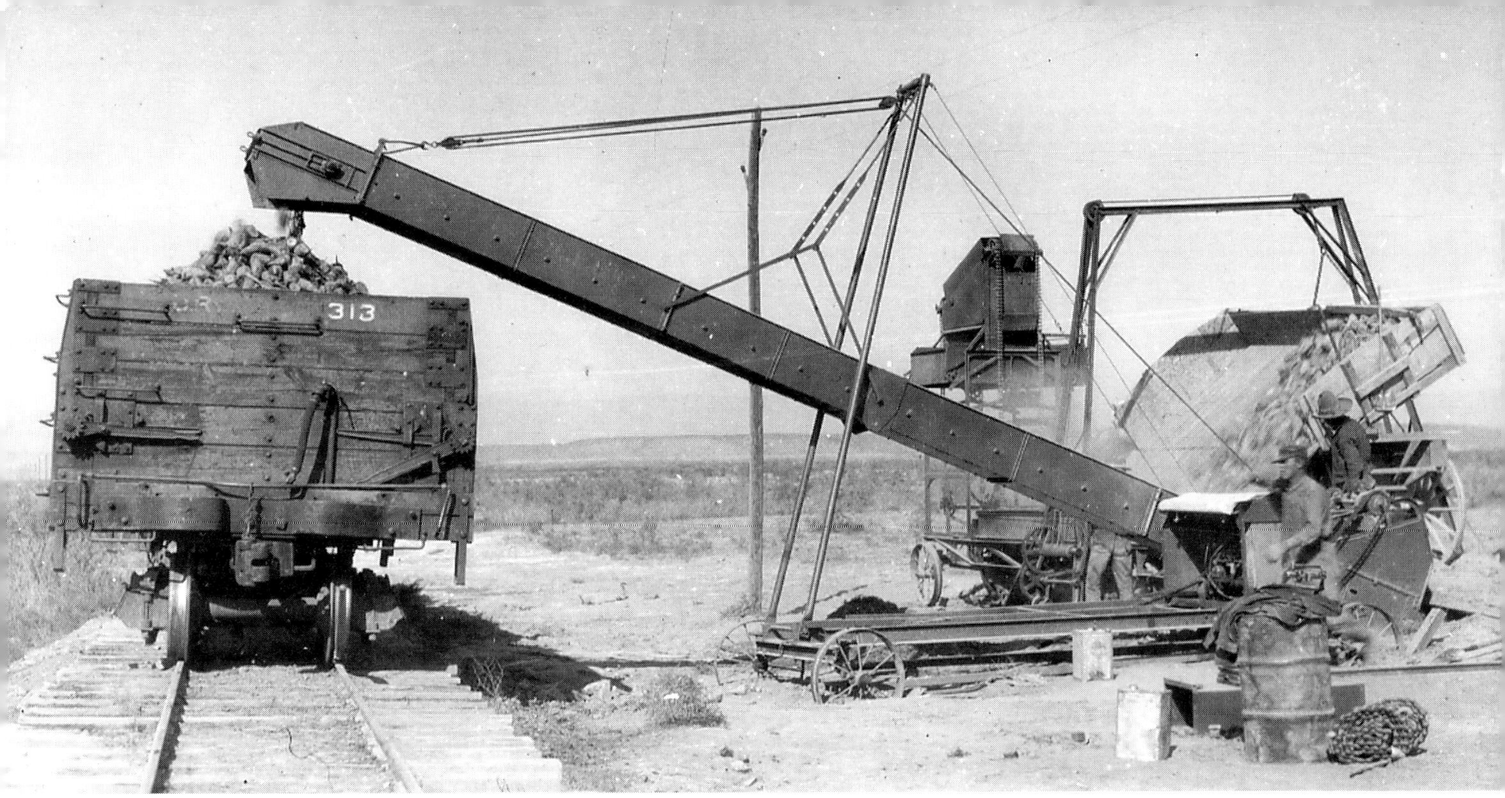

JIM McCOY COLLECTION

THIS CLOSE-UP VIEW shows the complexity of the early equipment used to load gondola No. 313. The sugar beets were hauled to the beet-dump area on horse-drawn wagons, which were driven under an A-frame. Cables were attached to one side of the box of the wagon, and a pulley system pulled it up on its side. The opposite side of the wagon swung open like a tailgate, and the beets were delivered to a conveyor belt that fed into the elevated conveyor. The self-powered elevator was raised and lowered on another pulley system controlled by the operator. In this way, the sugar beets were lifted to the height of the gondolas. Notice the empty gasoline cans and the roll of extra belting for the conveyor in the foreground.

THE UINTAH RAILWAY COMPANY

CAR CERTIFICATE

No. 1. ....STATION.................................192.....

I hereby certify that I have this day personally examined ....car No.....................and that the roof and sides have no loose boards, holes or cracks, or unprotected decayed spots liable to hold sparks and start a fire; that the king bolts or draft bolts are properly protected, and that there are no uncovered irons or nails projecting from the floor or sides of the car which might injure packages of explosives; also, that the floor is in good condition and has this day been cleanly swept before the car was loaded; that I have examined all the axle boxes and that they are properly covered, packed and oiled, and that the air-brakes and hand-brakes are in condition for service.

....
Railway Employe Inspecting Car

No. 2. ....STATION.................................192....

I hereby certify that I have this day personally examined the above car, that the floor is in good condition and has been cleanly swept, and that the roof and sides have no loose boards, holes, cracks or unprotected decayed spots liable to hold sparks and start a fire; that the king bolts and draft bolts are protected, and that there are no uncovered irons or nails projecting from the floor or sides of the car which might injure packages of explosives; that the explosives in this car have been loaded and stayed, and that the car has been placarded according to paragraphs 1661, 1666, and 1674 to 1683 inclusive, of the Regulations for the Transportation of Explosives prescribed by the Interstate Commerce Commission; that the doors fit or have been stripped so that sparks cannot get in at the joints or bottom.

....
Shipper or Duly Authorized Agent

....
Railway Employe Inspecting Loading and Staying

NOTE.—Both certificates must be signed. Certificate No. 1 by the representative of the carrier. For all shipments loaded by the shipper, he or his authorized agent, must sign certificate No. 2, and the representative of the carrier must certify as to loading or staying and general condition. When the car is not loaded by the shipper, certificate No. 2 must be signed only by the representative of the carrier. A shipper should decline to use a car not in proper condition.

OGER POLLEY COLLECTION

UINTAH ENGINE 10 was pushing a few gondolas into place next to the ditcher, to be filled with debris and dirt. This scene was photographed during June of 1926, a few months before No. 10 was sold to the Eureka–Nevada Railway. When the roadbed was undermined and washed out by the rains, it often was covered by mudslides and small washes. The loose shale would sluice down small hillsides and valleys on its way to West Salt Wash Creek, often damming up behind the Uintah's grade, causing the rails to be covered with mud. Although this did not cause an immediate emergency like a washout would, it had to be cleaned off the grade, because when it did finally dry out, the mud would become almost like concrete, causing an empty flatcar or caboose to leave the rails. The Uintah had hundreds of culverts along the right of way, but debris and slow-moving mud slides often caused them to "silt in" creating a large pond of water on that side of the tracks. Although the American Hoist & Ditcher was used primarily on the slides on Baxter Pass, it was used frequently in the desert between Mack and Carbonera. A small siding would be constructed to allow the ditcher to perform the work that it was designed for. Drainages along the track and mudslides near cuts and fills provided the ditcher and the section crews with an ample amount of work.

ROY ENO JR. COLLECTION

DAN PRICE COLLECTION

IN THE SUMMER OF 1924, several cars from a Uintah freight train heading toward Mack jumped the rails near Milepost 6.0. Stock car No. 411 and one of the steel water cars rolled off the track and tumbled to a stop in the flat desert. The stock car was nearly ripped in half, as momentum snapped the timbers all across the front end of the car. The water car ended up overturned and quickly lost its load of Mack's domestic water supply into the dry desert soil. By the time these two photographs were taken, the track had been cleared, and another train of flatcars loaded with bags of Gilsonite ore was heading past the wreck to Mack.

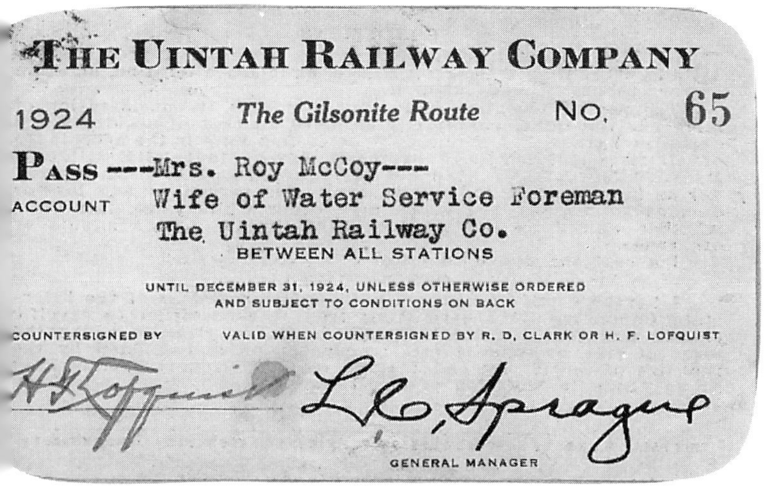
THE UINTAH RAILWAY COMPANY

1924 *The Gilsonite Route* NO. 65

PASS ---Mrs. Roy McCoy---

ACCOUNT Wife of Water Service Foreman
The Uintah Railway Co.
BETWEEN ALL STATIONS

UNTIL DECEMBER 31, 1924, UNLESS OTHERWISE ORDERED AND SUBJECT TO CONDITIONS ON BACK

COUNTERSIGNED BY — VALID WHEN COUNTERSIGNED BY R. D. CLARK OR H. F. LOFQUIST

H F Lofquist — L. C. Sprague
GENERAL MANAGER

RODGER POLLEY COLLECTION

THE UINTAH'S STRETCH of trackage from Milepost 5.0 to nearly Milepost 9.0 was an unbroken tangent—a good area for engineers to "make up some time." On at least two documented occasions, this increase in speed led to derailments. Both of these occurrences happened near Milepost 6.0, as each train obtained its fastest momentum, and before the brakes were applied too heavily. Bridges 6-A and 6-B were both short wooden trestles built for the drainages of West Salt Wash. The terrain around Milepost 7.0 probably was the most bland and uninteresting landscape on the entire line. The shale hills stretched away to the east and flattened, while West Salt Wash flowed away from the tracks, leaving the Uintah's roadbed in the middle of a sun-baked flat, Even the sagebrush was sparse at this desolate location. Near Milepost 8.0, the shale hills on both sides of West Salt Wash started to close back in, and the terrain next to the track became much more interesting in the next few miles. In 1929, a sign post was added at Milepost 8.0 in honor of Superintendent Walter Rader.

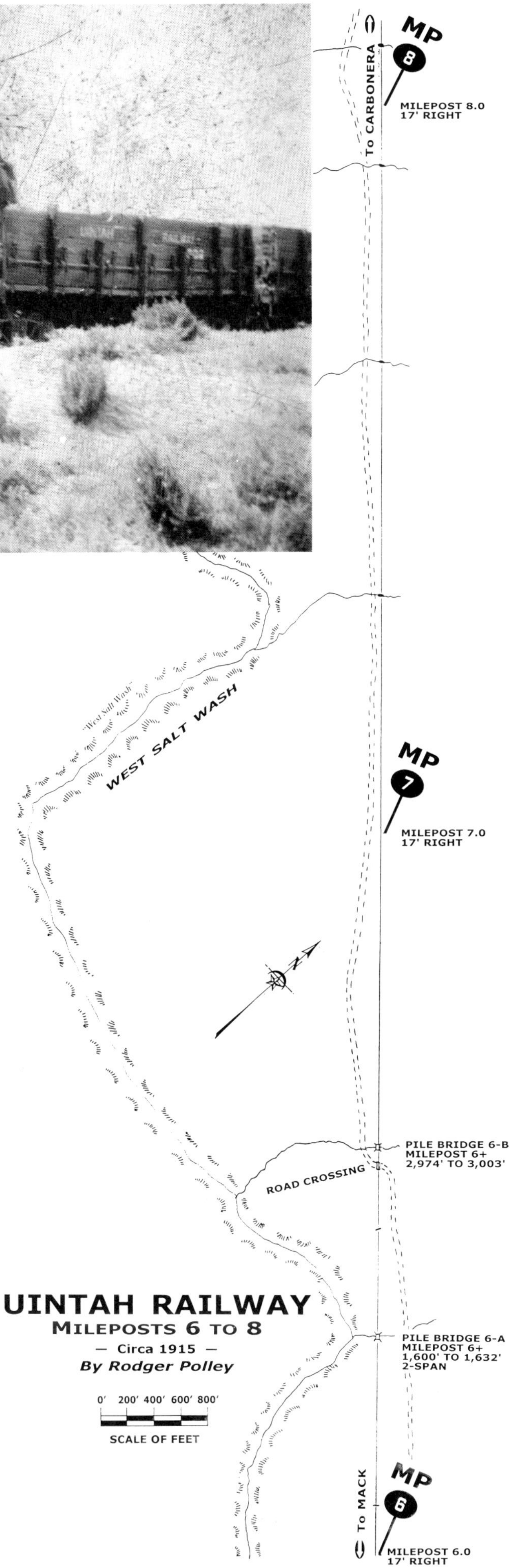

UINTAH RAILWAY
MILEPOSTS 6 TO 8
— Circa 1915 —
By Rodger Polley

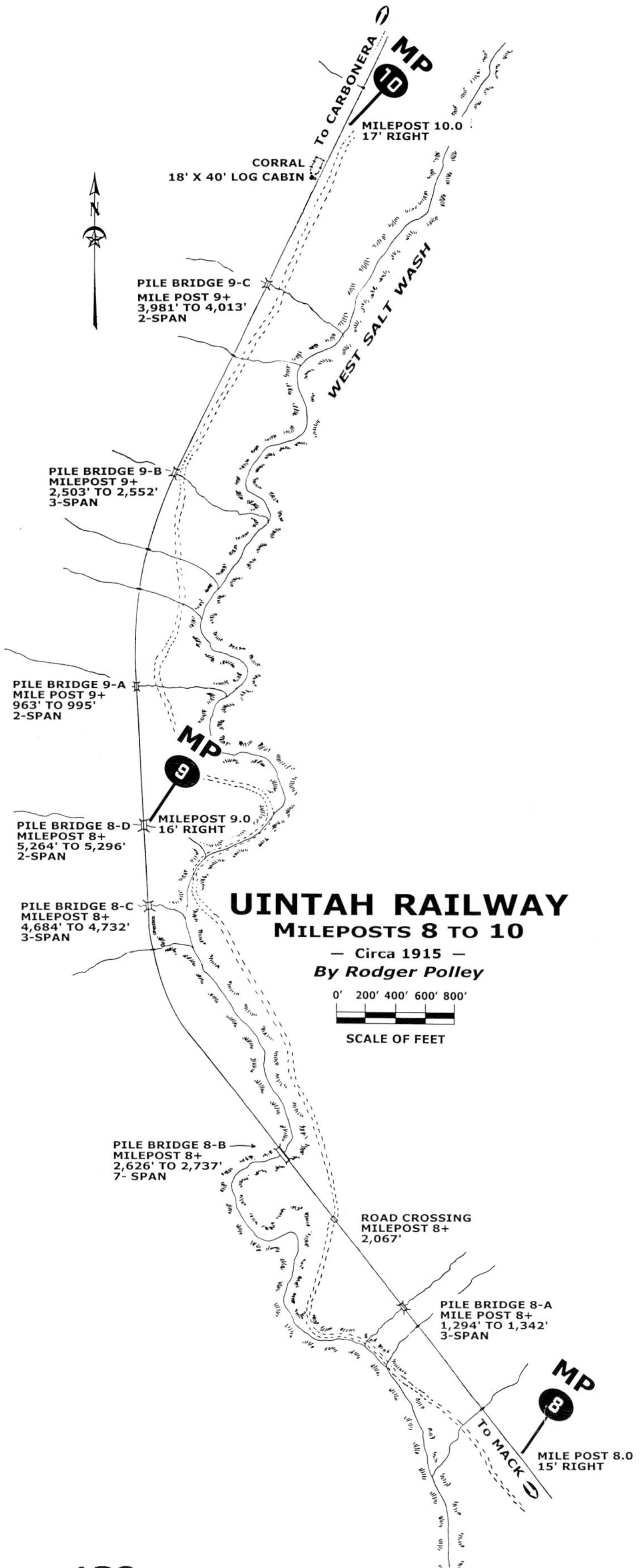

The terrain started to shove in closer to the track at Milepost 8.0, filling the horizon to the west and east of the line with the rough outline of eroded shale hills. The decomposed shale that made up the hills just west of Bridge 8-A was a perfect example of the bad conditions the Uintah had to contend with when the normally dry climate changed to rainy weather. This particular stretch of shale had very little vegetation at all, and when a severe thunderstorm came through the area, it was necessary for this trestle to be continually repaired or rebuilt, due to the massive amounts of mud and debris that literally eroded the dirt away from around the pilings in a single storm. At Bridge 8-B, the Uintah crossed West Salt Wash to the west side of the creek. Not far past the bridge, the track curved north along the wash, continuing toward the ever-closer Bookcliff Range. This area between Bridges 8-B and 8-D was a large and almost-continuous fill. The Uintah's track followed the bottom edge of the shale hillsides from Bridge 8-D to Milepost 9.0, crossing several small trestles and culverts used to cross the many drainages from these hills to West Salt Wash. Near Milepost 9.0, there was a large cabin and a sheep corral built close to the track, which provided sheepherders with a place to corral and ship their sheep to Mack or Watson.

Salt Wash — The section of track between Mileposts 10.0 and 11.0 was rather dull and mostly tangent. The wash was to the east of the track, and low-lying shale hills made up the horizon to the west. As the track curved at Milepost 11.0, the Bookcliff Mountains started to loom up in the distance. The entire northern horizon was made up of the large sandstone tiers that sit on top of one another, making what appears to be an impenetrable barrier. As a passenger on the Uintah gazed at the mountains ahead and wondered how their train could possibly surmount the obstacle, attention was brought back to the area under the train as it went over Bridge 11-A. This section of track between Bridge 11-A and Bridge 11-B was the site of a small section house known as Salt Wash, little more than a passing siding. In the 1920's, Salt Wash siding was renamed Sprague, after General Manager Lucius Sprague. A section house located here was used for housing a section crew working between Mack and Carbonera. However, sometime between 1915 and 1919, it was either torn down or burned down. This must have been a desolate place to live for any amount of time, because it was the most isolated and uninviting stop on the Uintah. Several miles from the farms at Mack — in the middle of the desert, with no shade and no water — this location was open both to bitter winter winds and hot summer sun. It should be noted that although West Salt Wash did run a trickle of water most of the year, it was just that — a trickle — and the water was full of alkali salt.

RODGER POLLEY COLLECTION

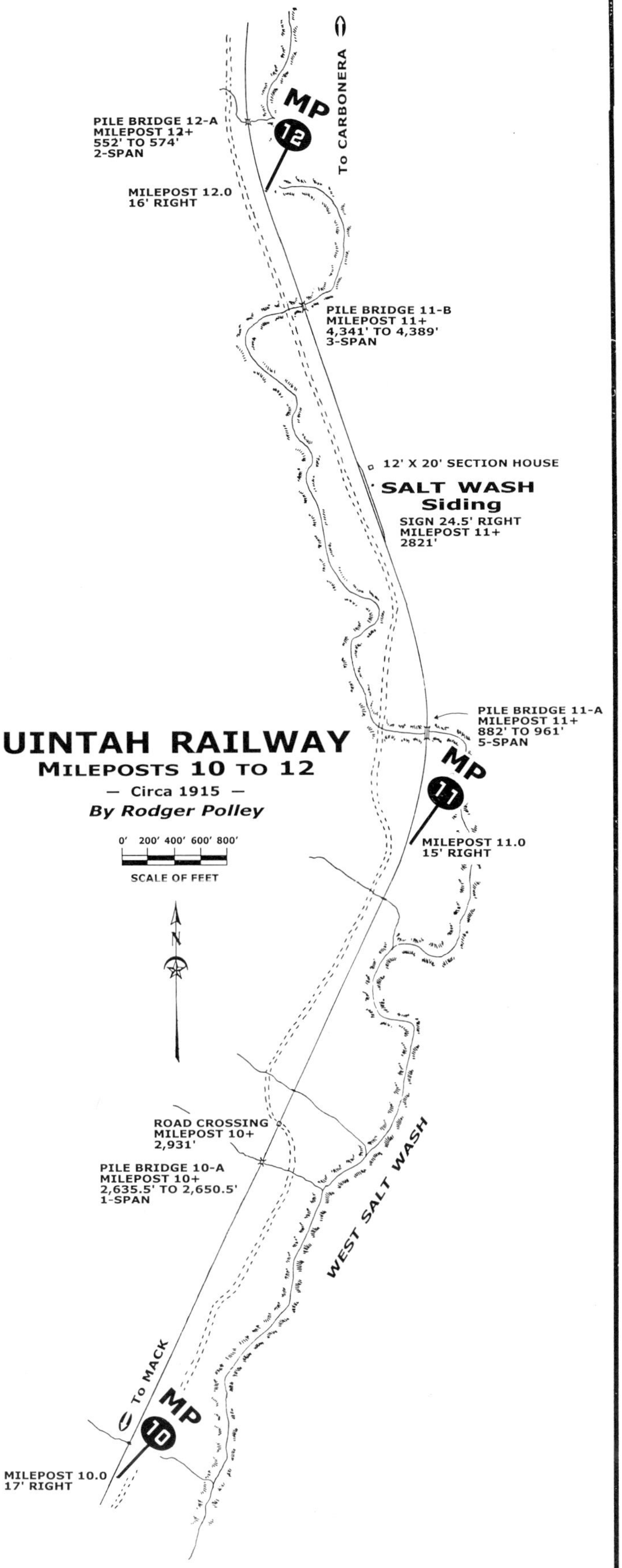

RODGER POLLEY COLLECTION

LOOK WHAT A FLOOD DONE was the handwritten inscription on the back of this photograph of Bridge 11-B. Taken from west of the track, looking across West Salt Wash, the flood waters had undermined several bents of the trestle, leaving it heavily bowed in the middle. Flash floods had this effect on many of the Uintah's wooden trestles, and one rainstorm left 12 bridges in disrepair from Mack to Atchee. The heavy erosion in this area is evident in the background, where the level of the valley had been washed away, leaving a high dirt bank towering above the creek below. To help keep the water from undermining the bridge abutment, the bridge crew embedded a row of iron stakes in the ground. Spare pieces of rod, pipe and rail were hammered into the soft soil in long rows, to help catch pieces of sagebrush and debris coming down the channel. This debris would then fill-in with dirt and sand, and if the rains did not last too long, this was usually enough to keep the abutments safe. The trestle bents in the middle of the stream usually were not so lucky. Even though the Uintah used the pile driver to push the bridge timbers deep into the ground, enough fast-running water could quickly carry the soil away, often leaving timbers hanging in midair.

LOOKING SOUTH toward Mack, this was Bridge 11-B, after a flash flood had swept pile timbers away—leaving the trestle deck bowed and hanging without support.

RICHARD A. RONZIO COLLECTION

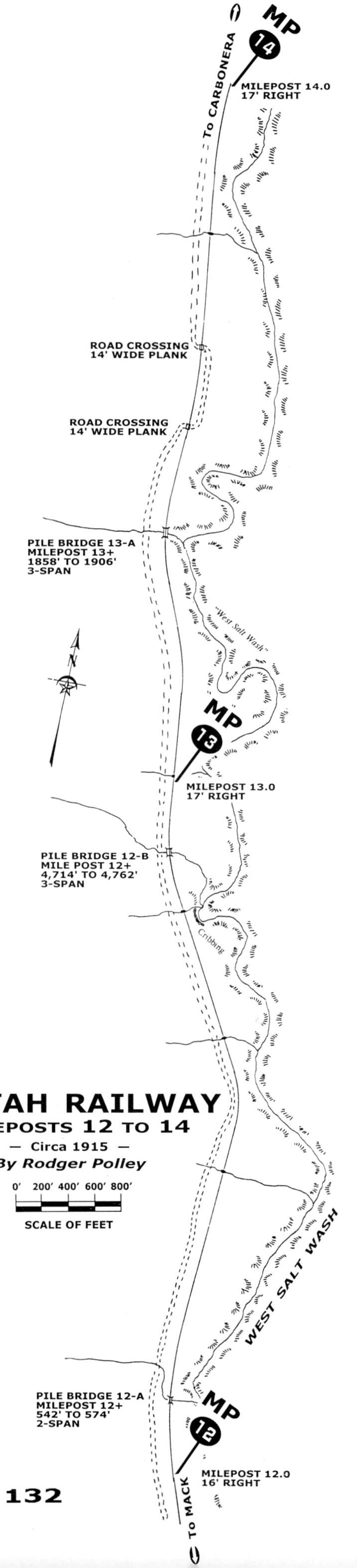

UINTAH RAILWAY
Mileposts 12 to 14
— Circa 1915 —
By Rodger Polley
0' 200' 400' 600' 800'
SCALE OF FEET

Between the Uintah's Mileposts 12.0 and 16.0, the line's trackage started to ascend into a totally different topography, compared with the low shale hills of the Grand Valley. Near Milepost 14.0, the Uintah's track was at the bottom of a large flat valley, and the horizon was starting to rise toward the sky in all directions. Mammoth shelves of sandstone, pushed up by geologic forces over time, filled the background with picturesque shapes. The force of erosion also added to the rugged look of the Bookcliff Range, as large chunks of sandstone broke away and rolled into the bottom of the canyon. A few cedar trees made these cliff faces their home, but for the most part, the cliffs were a barren outcropping of rock. Sagebrush and greasewood were fairly common all along the Uintah's right-of-way; however, as the tracks continued up the canyon toward Atchee, the plant growth became increasingly large, to the point of being taller than a man's head in some places. The minimal grade from Mileposts 14.0 to 16.0, on the flat valley floor, made it easy to construct roadbed and lay track. This area also saw a lot more rock and natural gravel in the soil, which made the roadbed less likely to wash away, as it frequently did in the soft shale, a few miles away.

RODGER POLLEY COLLECTION

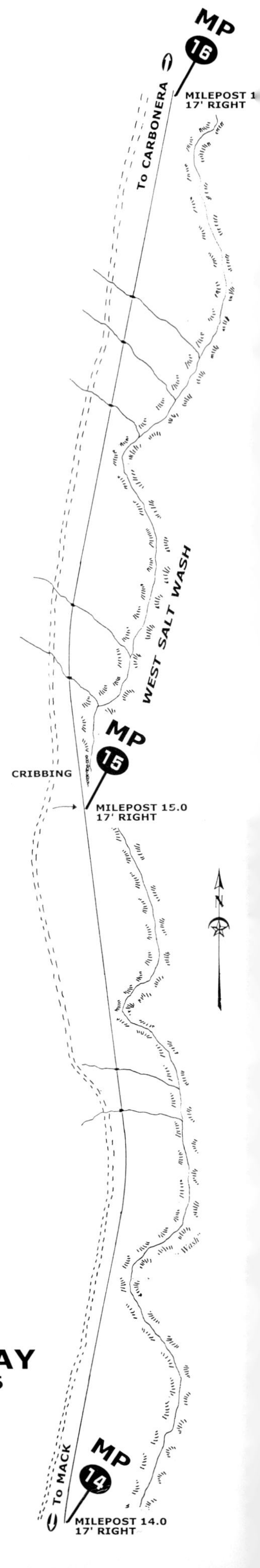

UINTAH RAILWAY
Mileposts 14 to 16
— Circa 1915 —
By Rodger Polley
0' 200' 400' 600' 800'
SCALE OF FEET

FRANK A. KENNEDY PHOTO
RODGER POLLEY COLLECTION

THE LARGE ROCK formation near Milepost 17.0 was named "The Crows' Nest," and was typical of the topography of the Bookcliff Range. Large layers of sandstone rise up from the softer shales below and are much more resistant to the forces of erosion. This picture of Railcar 50 was taken on a summer day, at the rock formation northwest of the track. The canopy was positioned on the railcar to allow the passengers a break from the hot summer sun. The Uintah's tracks breached the edge of the Bookcliff Range at this point, following the grade of West Salt Wash Creek. Cliffs, like the one in the picture, started to close-in at this point into the narrow sandstone canyon, carved by the creek.

DELL A. McCOY PHOTO

THE CROWS NEST was photographed on October 29, 1998, as the author and publisher drove from Mack to Baxter Pass. The rock formation was viewed as they looked toward the northwest, from the former Uintah Railway grade. Jet aircraft vapor trails obscured part of the clear blue sky in the west, creating man-made clouds just below the stratosphere.

MUSEUM OF WESTERN COLORADO – RESEARCH CENTER
BANKS COLLECTION

CARBONERA WAS IN its infancy during 1904, as the construction of the Uintah Railway was taking place. The log "duplex" was the first building in town, with the large frame houses being built later that year or in 1905. The large white tents were put up as temporary shelters for the contractors on the railway and for the miners. Just like any start-up operation, the emphasis was on production and efficiency, while the comforts of home were an afterthought.

RICHARD A. RONZIO PHOTO

RICHARD A. RONZIO visited Carbonera, Colorado, during the summer of 1959, while he was researching for a projected book on the Uintah Railway. Concerete foundations were still in place where the former coal tipple had stood, west of the railway track.

AN EARLY SPRING DAY in Carbonera during the teens was chosen for most of the town residents to line up for a photograph. The miner in the center of the picture was Hamp Bennett, and the miner with the light on his miner's hat was his brother, Peter Bennett. Their father and mine foreman, James Bennett, was on the far right of the picture. As a direct extension of the Gilson Asphalt Company, Mr. Bennett reported directly to them, and all business arrangements were totally separate from the railroad. Production was always tied to the Uintah management's decisions, but for everyday operations of the mine and the settlement of Carbonera, James Bennett had the final word for 30 years.

PETER BENNETT COLLECTION

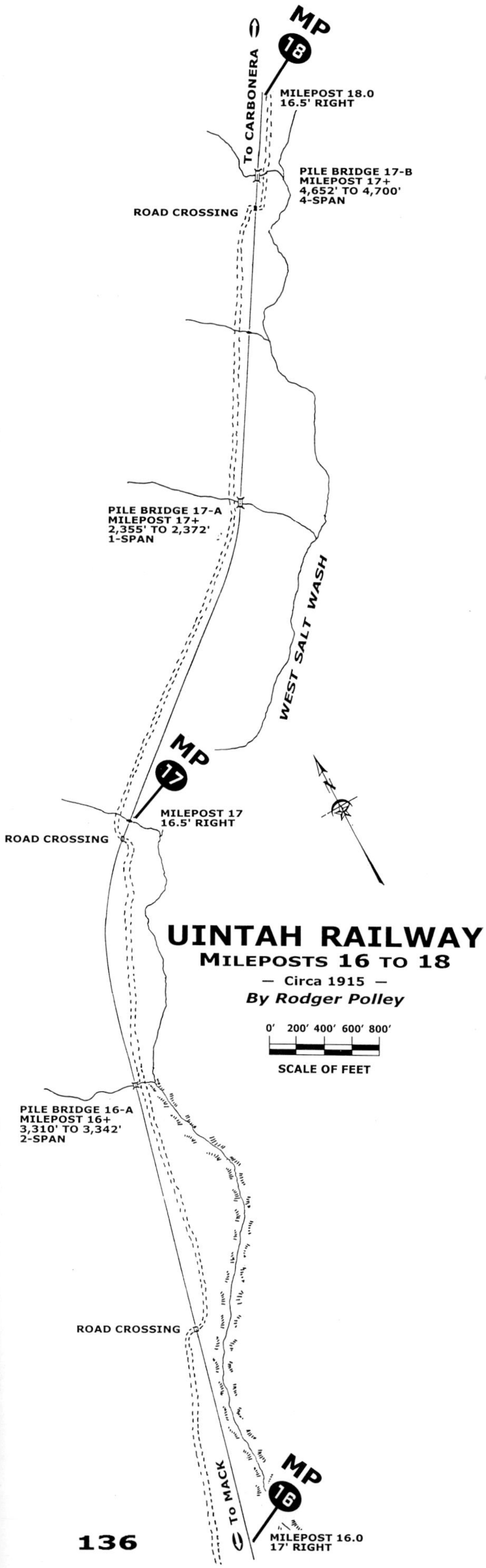

MILEPOSTS 16.0 to 20.0 saw more of the same terrain, as the Uintah's track was extended farther into a narrow valley made by West Salt Wash Creek. The trackage stayed on the west side of the creek and utilized small bridges to cross other "dry washes," which entered the main canyon from the side. The track slowly made its way back from a northeastern direction to northwestern, as it turned toward Carbonera. It should be noted that this was usually the area where most automobile travel was forced to halt during the winter months. The road to Carbonera and Atchee was not maintained, and for the most part, it was just two ruts through the sagebrush. Near Milepost 18.0, the higher elevation in the canyon provided a cooler atmosphere in the winter than in the Grand Valley desert. Motorists often became stranded in snowstorms as they drove toward Baxter Pass. Upon leaving Mack they would encounter very little snow, making them think the drive to the pass would be easy. This caused many problems for motorists who ventured up into the canyon. This phenomenon was not only due to the canyon, but also was because of the ever-increasing elevation, as the Uintah's track wound its way upgrade. Colder temperatures during winter months were almost a definite circumstance the closer you got to Baxter Pass. By the time you reached Atchee, much deeper snow had fallen and accumulated than in Mack.

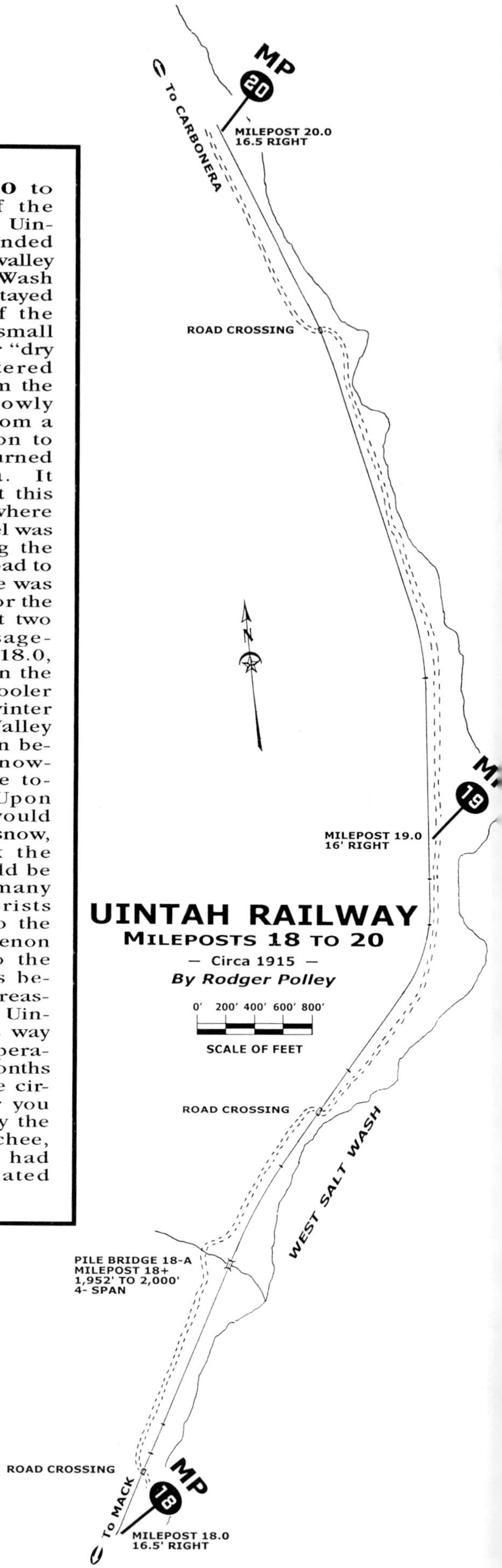

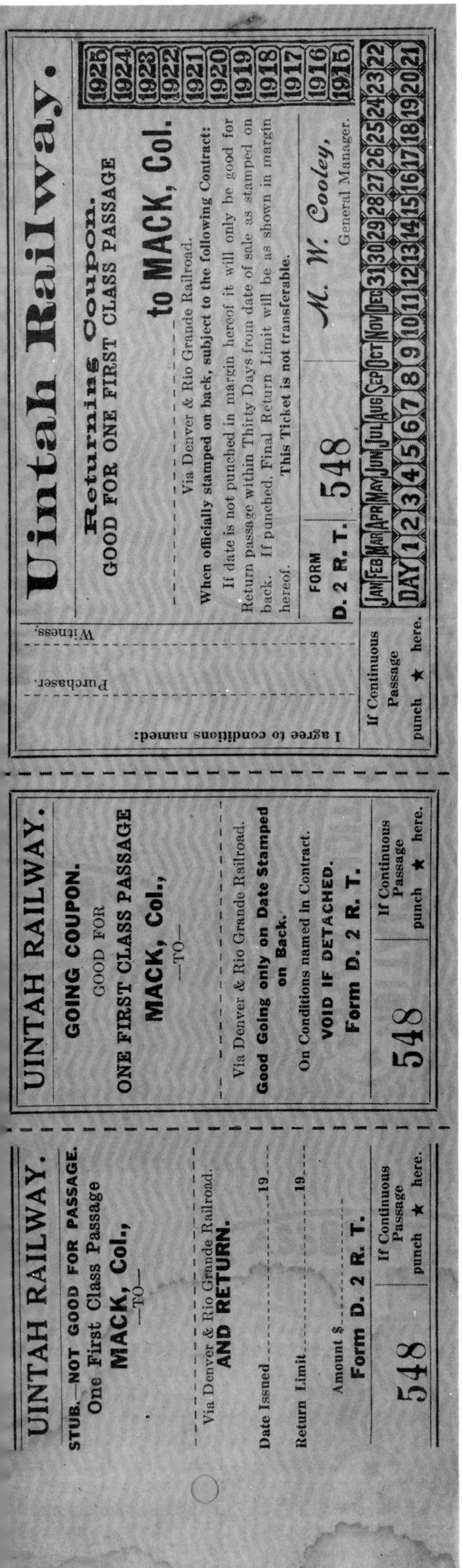
UINTAH RAILWAY.
STUB.—NOT GOOD FOR PASSAGE.
One First Class Passage
MACK, Col.,
—TO—
Via Denver & Rio Grande Railroad.
AND RETURN.
Date Issued..................19.....
Return Limit..................19.....
Amount $..........
Form D. 2 R. T.
548
If Continuous Passage punch ★ here.

UINTAH RAILWAY.
GOING COUPON.
GOOD FOR
ONE FIRST CLASS PASSAGE
MACK, Col.,
—TO—
Via Denver & Rio Grande Railroad.
Good Going only on Date Stamped on Back.
On Conditions named in Contract.
VOID IF DETACHED.
Form D. 2 R. T.
548
If Continuous Passage punch ★ here.

Uintah Railway.
Returning Coupon.
GOOD FOR ONE FIRST CLASS PASSAGE
to MACK, Col.
Via Denver & Rio Grande Railroad.
When officially stamped on back, subject to the following Contract:
If date is not punched in margin hereof it will only be good for Return passage within Thirty Days from date of sale as stamped on back. If punched, Final Return Limit will be as shown in margin hereof. This Ticket is not transferable.
FORM D. 2 R. T. 548
M. W. Cooley, General Manager.
1915 1916 1917 1918 1919 1920 1921 1922 1923 1924 1925
JAN FEB MAR APR MAY JUN JUL AUG SEP OCT NOV DEC
DAY 1 2 3 4 5 6 7 8 9 10 11 12 13 14 15 16 17 18 19 20 21 22 23 24 25 26 27 28 29 30 31
I agree to conditions named:
Purchaser.
Witness.
If Continuous Passage punch ★ here.

CHRIS H. UEBERROTH COLLECTION

COLORADO HISTORICAL SOCIETY COLLECTION

ON OCTOBER 25, 1908, the Uintah Railway experienced one of its first accidents resulting in the serious injury of an employee. Passenger engine No. 21 was running back to Mack from Atchee, and as had been the case that month, there was a lot of rain that day. A large rainstorm had sent a flash flood of mud and debris flowing down the side of the hills west of the track, somewhere in the canyon between Mileposts 18.0 and 20.0. The mud and rocks were quickly deposited between the rails, and the water continued over the side of the track, undermining the roadbed on its descent to the bottom of the wash. The soft ground quickly gave way under the weight of the engine, causing her to overturn onto the left side —nearly rolling all the way over on the incline of the roadbed. Fireman Roy Miller was caught off guard, and his foot was quickly scalded. The engineer escaped serious injury, but Roy Miller would had to have his toes amputated because of the scalding. This view shows the remains of the wreck, after the clean up was initiated. Notice the top of a dome, off the boiler on the ground, at the right. Large chains and a cable had been thrown over the top of the engine, while men dug a hole beneath it to access the other end of the cable. Once attached, and the ground prepared, the engine was righted, using "come alongs" and then pulled up to the track level. Engine No. 21 would be repaired at Atchee. A few weeks later, folk who had seen Roy Miller in Fruita, reported that he was in "good spirits" and looking forward to getting back to work.

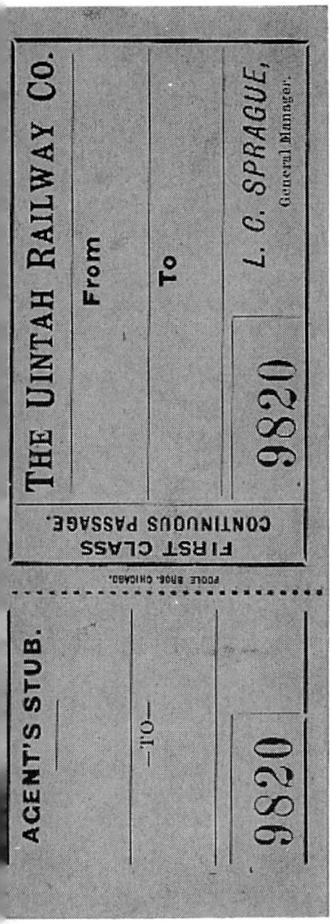
THE UINTAH RAILWAY CO.
From
To
L. C. SPRAGUE, General Manager.
9820
FIRST CLASS CONTINUOUS PASSAGE.
POOLE BROS. CHICAGO.
AGENT'S STUB.
—TO—
9820

ODGER POLLEY COLLECTION

OTTO C. PERRY PHOTO – RICHARD H. KINDIG COLLECTION

UINTAH LOCOMOTIVE No. 50 was heading northbound on May 9, 1939. This portion of the Uintah's route started to get much more scenic than it was in the flat shale hills to the south, toward Mack. The large sand-

stone ridge makes for some very scenic views in the desert country for the next several miles. This photograph was taken as Otto Perry was looking southwest, near Milepost 18.0.

Text continued from Page 120...
out of the entrance of the mine and came to a switch. The switch was always turned away from the incline for safety. The tracks ran across the small gulch and slowly went uphill to a wooden buffer at the end of the track. This made sure that if a coal car got away, it would run past the incline and come to a stop on the siding. A horse was usually employed at the entrance to the mine to switch the tram cars out. In the 1920's, a large white horse was used and remembered by many who lived in Carbonera. The horse and mules were usually kept in a small corral near the section house.

The entrance to the mine was dug into the hill and directly into the coal seam. This original tunnel, which went back into the mountain in a southwestern direction, was not improved and had sandstone and dirt walls at the entrance. The mine track went up the center of the corridor and, for the first 50 yards or so, ran at a slight incline according to the pitch of the sedimentary rock that the coal seam was in. Then, at that point, the track went slightly downhill, back into the mountain at a rate that was barely discernible. Not far back in the mine, an original air shaft was cut back east toward town and exited the mountain in the canyon right behind town. This small shaft was a favorite place for children to play in and around, and it could be used as an escape if a cave-in happened near the main entrance. Mining at the face of the mine was done with hand picks. No pneumatic tools were ever used in Carbonera, owing to the low production that was needed.

ENGINE No. 40 was sitting on the mainline, next to the siding at Carbonera. The stub trackage to the right entered a small shed where a section gang's motorcar was stationed. The piece of rail lying perpendicular across the top of this storage track was used as a safety devise to keep their section car from rolling onto the track. No. 40 was familiar to these tracks, often running extra trains to Carbonera to pick up loads of coal, for both Mack and Atchee.

DENVER PUBLIC LIBRARY, WESTERN HISTORY COLLECTION

OTTO C. PERRY PHOTO – RICHARD H. KINDIG COLLECTION

THIS SCENE PORTRAYS Engine 40 as she began to pull out of Carbonera on September 3, 1934, heading north toward Atchee with a train of coal and empty flatcars, with combine No. 50 at the end of the consist. Carbonera's platform for passengers and parcels was in a rather desolate location. One lone person stood next to the track at this "station platform." Located on the mainline and not on the siding, passengers, express and small L-C-L (less-than-carlot) shipments were let off and picked up at a small wooden platform on the west side of the track, far to the south of the siding and most of the town. Its location probably was due to being closer to the frame company houses, which were at the south end of town. Perhaps even more likely, it was due to the fact that the mine foreman lived in one of these houses!

RODGER POLLEY COLLECTION

TAKEN FROM THE TRAIN, this picture shows nearly all of the sprawling town of Carbonera. There were just enough houses for the employees who worked there, as well as a railway section house, ice house, tool shed and some chicken coops, which made up the entire town. At the head end of the train, the passing siding headed toward the loading tipple. The section house sat next to the only tree in town, while the rest of the dwellings were lined-up parallel to the Uintah's mainline. The two "tar-papered" bunkhouses closest to the camera each had three small bedrooms and were used by miners for sleeping quarters. These two bunkhouses and another one, located closer to the tipple, were primarily overseen by the mine superintendent, James B. Bennett and his wife, Sarah. These two people were virtually "fixtures" in Carbonera, nearly from its inception until the Uintah pulled out in 1939.

Miners were paid by different rates over the life of the mine, from daily, hourly and by production. No matter what the pay scale, an eye was kept on individual production. In the 1920's and 1930's, this created a weighing system that allowed the miners to be paid by the tonnage. Small brass plaques similar to dog tags with numbers on them were used to keep track of a miner's tonnage. A number would be assigned to an individual and he would pick up his "chips" in the morning at the small scale house on the tipple. At the face of the mine, each miner would load a single tram car at a time and hang his brass tag on the car. The coal car was then taken out of the mine, let down the incline, and weighed on the scales at the

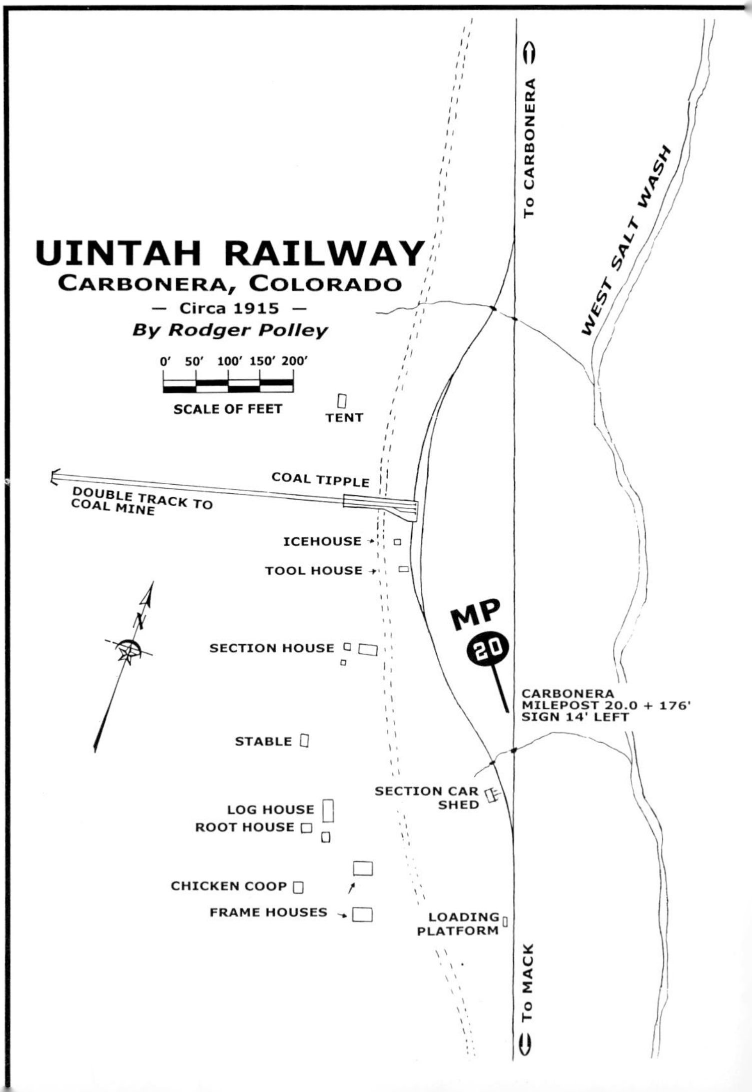

RODGER POLLEY COLLECTION

LAUNDRY WAS HANGING from the front of Harold and Juanita Workman's white frame boarding house. It had a large kitchen and dining area, where the family who lived in the boarding house prepared three meals a day for the mining crew. This house also was the location of all the food preparation for the miners during the fire in 1934, when nearly 100 miners and rescue workers made Carbonera their home for a few days. Behind the Bennett's home, the family kept a chicken house, which can be seen next to the small "privy." The house in the center of the picture, with the screened-in porch, was that of the mine foreman, Mr. Bennett. The log building at the right was a small apartment of sorts, with two small two-room living quarters at each end. The large outcropping of sandstone over the top of the log house marks the area in the canyon beyond the point where the mine had a small ventilation shaft that exited from the side of the mountain. This is where the young boys in town found a nice cool place to play during the summer, and where a child could scurry back into the mine to retrieve a certain tool or deliver a cold lunch for dad.

MAP AT CARBONERA...

CARBONERA WAS ESTABLISHED before 1915. This was before the Uintah Railway's engine roster and traffic load increased, resulting in more miners being hired, and in turn, more bunkhouses were built south of the frame houses. The buildings shown were built by the railway and torn down when the railway was abandoned. The small building between the "root house" and the "log house" was the bathhouse, used by all of the miners to "wash up" before going to their homes. An addition was built onto the front of the section house sometime after 1915, including a screened porch. Ice from Lake McAndrews and water from Atchee were delivered here and used by all the families. The tent north of the tipple was a temporary living site, as there are no pictures of that area showing anything but sagebrush. The original map does not show the mine entrance or layout; however, in scale, they would have been at the end of the two tracks drawn on this map.

RODGER POLLEY COLLECTION

tipple. The mine foreman would then keep track of the amount of coal each miner had mined and record it in the pay schedule.

The inside of the mine was shored up by using timbers supplied by the railroad. Local production of timbers for the mine was done on the north side of Baxter Pass early in the Uintah's history, but by the mid-teens, the wood timbers were brought into Mack and shipped to Carbonera and to the Gilsonite mines in Utah on the Uintah Railway. The timbers were loaded into the coal-mine tram cars and brought up to the mine on the incline and then taken back into the mine with the mules. Since the tram cars had no brake rigging, much care was taken when the cars were not on a level surface. Oak sticks (?) were used to lodge into the wheels to stop the cars. If a car, especially a full one, picked up too much speed, it could easily lead to an accident.

During the early 1930's, Harold Workman, a long-time employee of the mine, started his tram car down the slight incline just inside the entrance to the mine. Riding on the front of the full tram car as he had many time before, he picked up speed until he decided to reach down and put the oak stick into the wheels to

PAUL AND BETTY GARBER COLLECTION

COMPANY HOUSE No. 4 was one of two frame houses divided into apartments for mining crews during the early years. This picture of the front porch area was taken during the first few years of the Uintah's operations. Later, the coal consumption of the railway required more output from the mine, and in turn, more miners arrived, and additional houses were built in Carbonera. After World War I, the two frame houses were converted into an eating house for the miners and a single-family dwelling for the mine foreman. Three small bunkhouses, designed to house several men each, were built for sleeping quarters. Life in Carbonera was not a picnic by most standards. The scissors on the wall were used for haircuts on the front porch, and in the background, a wash tub on the side of a small log shack allowed a miner to get a tub full of water for washing. Carbonera had no electricity or running water, and all the water in town was brought there by the railway, deposited in the cistern, and then hand-pumped by the user into a bucket to be utilized as needed. On top of all this, there were very few women who appreciated this poor standard of living. Most of the miners had families that lived in Mack, Fruita or Grand Junction, and trips on the weekends were the only time the miners saw their wives and families.

slow it down. Hanging out the side of the car, Workman misjudged the distance to the entrance doors to the mine and the distance between the car and the entrance way. In an instant Workman was pinned between the heavy tram car and the entrance-way posts. The little tram car rolled on past, and Workman was left bruised and with his right hip crushed and nearly sheared off. Workman ended up being in a cast from his chest on down the length of one leg for nearly a year. The prognosis for his recovery was pretty grim; however, within one year he was back at the mine setting monthly records for individual production among the miners.

The large doorway at the entrance that Workman had hit had been built in the early 1930's in an upgrade of the mine for safety. Dangerous "blackdamp" gas, as it was known by miners, had started to show up in the mine at Carbonera as the mine workings continued deeper into the mountain. This bad gas

THIS SIDING SERVED the coal tipple at Carbonera. The Uintah's mainline is out of sight, to the right of this view. The building in the foreground was used to house two hand cars and a small motorized section car. The nearest turnout was used to switch cars to either the passing siding (on the right) or to the main loading track (on the left). Empty cars generally were left on the Atchee side of the tipple (upgrade), and then, the brakes were released so the cars could go down to the tipple by hand as needed. After loading, the cars were let go on down to the track south of the tipple. Just left of the loaded gondolas in this view was a switchstand that was used for a "derail," in case a car might become a runaway. This way, the runaway car could make it only to the derail devise, rather than continuing down to the mainline.

BOB BOULTER COLLECTION

RODGER POLLEY COLLECTION

THE PASSENGER "DEPOT" in Carbonera was located in the Uintah's section house, as seen here in a winter scene, with the CARBONERA station sign just above the front porch roof line. This was the residence of the section crew foreman, George Komata, who lived there until the Uintah Railway shut down. He worked with the section gang between Mack and Atchee for many years. The corrals where the mules and horses for the mine were kept can be seen to the left of the section house, and Mr. Komatas's cottonwood tree, complete with fence, can be seen at the right. The only potable water in town was located in a large cistern, just behind the gondola at the right.

RICHARD A. RONZIO COLLECTION

A CLOSER VIEW of the tipple reveals the spacing of the timbers above the cars on the inside track. These timbers were necessary to properly brace the outside bent for the thousands of tons of coal that would roll over the top of it for many years. However, their height was always an annoyance to railroaders, as no through access for the engines was provided on that track, and only "coal cars" could be run under the timbers. All cars had to be pushed through from one side or another, or in most cases, they were rolled downgrade with the help of hand brakes by those loading the coal. Notice the dirt road that goes under the tipple, at the left.

contained methane and could lead to anything from light-headedness and headaches to unconsciousness and explosions. To get fresh air into the mine, the Colorado state mine inspector required the Gilson Asphalt Company to build a ventilation system into the mine. Instead of a complicated and expensive system being built, a relatively inexpensive design was built. A second entrance was carved into the mountain just a few feet to the right of the main entrance. The tunnel was mined parallel to the main entrance for several yards and then brought back into the main tunnel. In the main tunnel, a large iron door was constructed near the entrance. A large fan, nearly 10 feet in diameter, was then installed in the new tunnel, and a wooden building holding the engine to turn the fan was built just outside the new tunnel. With the large fan turned on and the large iron door closed in the entrance shaft, fresh air was pumped into the mine.

The mine increased its production considerably in 1905, as the Uintah increased its locomotive roster. This would be the way of things for the next 20 years, as the production at the Gilsonite mines in Utah increased — and so did the coal production at Carbonera to supply the locomotives. With the extension to Watson and the new mines at Rainbow in 1911, the need for coal increased again, as the new mining towns expanded and even more miners and employees for the railroad were hired. For the next 16 years, the Uintah's fleet of locomotives would be between 11 and 14 engines, and the corresponding coal production at Carbonera made the late teens and 1920's the busiest times that Carbonera would ever see. Also during this time period, the boilers and electric plants at Mack and Atchee were increased in size, causing even more demand for coal. Added labor was needed at the mines. A small bunkhouse was built north of the log duplex, and later three small three-room bunkhouses were built to provide room and shelter to the miners. One of these bunkhouses was built between the log duplex and the section house while the other two were built south of the two frame houses. These were the last buildings built in Carbonera, and rounded out the look of the town — one row of small houses sitting parallel to the Uintah's tracks.

On December 6, 1919, the railway experienced a fatal accident at Carbonera. Conductor George Rucker was killed on the mainline near Carbonera. A malfunctioning coupler on the front of Shay No. 3 was to blame. One of the rod engines was to be coupled to the front of the Shay and its train when the front coupler failed. Conductor Rucker and Engineer Earp

THIS OUT-OF-FOCUS SHOT was taken from the north side of the tipple, looking back at Carbonera. Several loaded gondolas had been rolled downgrade (south), while one more empty car had been spotted beneath the tipple. The last empty gondola waited near the stock-loading chute. Loaded cars of coal then were switched off the lower holding track and assembled into a train. The ruling grade on this part of the line was 1.93 percent, and cars could be moved south on the siding by hand, using their hand brakes to slow them down. The small loading chute for livestock, at the right, was used by a few ranchers at this location, as well as for the mine's infrequent need for replacement mules.

RODGER POLLEY COLLECTION

DAN PRICE COLLECTION

ANOTHER PHOTOGRAPH of the north side of the tipple at Carbonera shows large piles of extra mine timbers that had been set off next to the tipple. This view was produced during the late 1920's. The scale shed on the tipple was just above the pile of timbers at the left. This small shed had a metal roof and tarpaper siding, but had no windows. The covered stairway had been constructed with different pieces of scrap lumber, leaving the outside with a striped image. The low-lying building, just past the tipple, was Carbonera's ice house.

ARTHER HILLMAN COLLECTION

THE FIRST No. 11 locomotive was on the siding at Carbonera, setting out empty gondolas, and had paused while the engineer checked the drivers and brakes. The dirt road that went from Carbonera to Atchee passed under the tipple through the fourth section from the concrete retaining wall. As built, this originally was the only section that was not cross-braced, but later, for whatever reason, the bracing was removed from the rest of the tipple.

A SMALL STRING of gondolas had been left in front of the tipple, waiting to be filled from above. Two tracks ran to the tipple, one going under the last section and one going right in front of it. Here loading is taking place on both tracks. Notice the extra coal that has piled off of the gondola onto the ground next to the tracks. Empty cars were moved forward for loading by gravity.

ARTHER HILLMAN COLLECTION

were working on the problem, when all at once enough steam built up in the waiting locomotive to cause it to lurch backwards, catching the unsuspecting conductor between the two locomotives. Mr. Rucker died on the spot, being survived by his wife and a son who later worked on the Uintah as a section man.

With the Uintah Railway's purchase of the two new articulated engines in 1926 and 1928, and the subsequent sell-off of extra motive power, Carbonera's mine saw its first slump in production. As the ever-efficient Nos. 50 and 51 took over more and more duties as motive power, and as the Gilsonite mines slowly lowered their production due to the Great Depression of the 1930's, even fewer miners were needed to produce the coal. By 1933, one of the bunkhouses was turned into a small school for the six to eight children who were growing up in Carbonera at the time. The school only lasted a few years, and when the railway was abandoned, the school was closed down.

Flash floods were the most common source of concern for the people at Carbonera. This particular part of the canyon was wide and flat and full of sand. Large "gully washers" would snake their way down the narrow valley above and spill out into the wide portion of the valley. Walls of water seven or eight feet high were common in some years. Although this did provide an easy method of getting rid of the town's trash, on several occasions it was taken much more seriously as the water swept over the Uintah's roadbed, leaving large pools of water standing in the low areas in front of the house and section house.

The worst disaster that affected Carbonera was not natural, but man made. On Easter Sunday, April 1, 1934, a fire broke out in the coal mine, killing one miner and putting a halt to activity on the railway. It was mid-morning when John Rogers, a 51-year-old miner, whose family lived in Palisade, Colorado, decided to put in a little extra time working in the mine. Paid by production, his family at home, and a lot of the rest of the town in Grand Junction for the weekend, Rogers and his dog climbed the incline to the entrance of the mine. He entered the enginehouse to start the fan, and in no time, he was walking back to the face of the mine, surrounded by fresh clean air from outside. The large gas engine that propelled the fan was water-cooled, and care was needed to make sure that sufficient water was in the tank that sat above it. It is not known if the engine ran out of water or was just low on oil, but for whatever reason, the engine became very hot and ignited the wooden building around it. Within a few seconds, the large fan

RICHARD A. RONZIO COLLECTION

MINE FOREMAN and manager, James Bennet, was standing next to one of the large mine-tram cars, which reportedly held about a ton of coal. The track on the left went out to the end of the tipple to a wooden chute, while the one at the right ended at a chute in the middle of the tipple. This chute went down to the Uintah tracks beneath it. Scales for weighing the coal were kept in the tar-paper shack, at the right.

CARBONERA'S TIPPLE had three coal chutes to deliver coal to the empty cars spotted on the siding. Over the years, the location of these chutes was moved from the front tracks to the tracks under the tipple. It is unclear if this was designed into the tipple, allowing the foreman to change it on demand, or if this was a modification produced over time. A covered stairwell went up on the right side of the tipple to the deck, and it was used so workers could easily get from track level to the deck of the tipple. Regularly scheduled carloads of coal were handled by the miners, with help from the train crew to load gondolas. However, when coal was needed just for the engine, the train crew often filled the tender by themselves. You can make out two of the mine's tram cars, passing one another near the mid-point on the incline. At the right, beyond the gondola, a pile of timbers were stacked, waiting for a trip to the top of the incline and on into the mine to be used to shore up the ceiling in the mine.

GEORGE ROSELAND COLLECTION

RICHARD A. RONZIO COLLECTION

COAL MINED AT CARBONERA was shoveled directly into tram cars and then was moved to the entrance of the mine by mules to be set out on a small siding. A large horse was used outside the mine to switch tram cars at the top of the incline. The weight of the loaded cars pulled empty cars up from the bottom of the hill, while a large brake mechanism was utilized to slow the descent. Also, part of the same mechanism was a hand crank, which was geared to allow someone to crank an empty car up to the top—although this was infrequent and an unpopular task. A simple switch made of a single piece of rail was hinged at one end and swung from side-to-side, as the car's weight crossed the switch. Farther down the tracks, there was another unusual area in the track, in the form of a small curve in the rails, just above the switch. Whether or not this was designed into the track, or was merely a construction oversight, is unclear. The rails split at the switch, and on the left side, split again at the end of the trestle. As can be seen, the place where the tram cars were brought to a stop often allowed some coal from the load to be thrown out on the ground. Later, a more complex system of switching was installed in the area where the tram car was sitting. A small shack was on the right side of the trestle, and scales had been installed to weigh the coal as it came onto the trestle. A crossover track allowed the coal cars to be switched to the left and right tracks more easily. Instead of having to go up the incline to the switch, cars could be moved around on the trestle by hand on the level. This picture was taken early in Carbonera's history, and the wooden tram cars eventually were replaced with all-steel models.

PETER F. BENNETT COLLECTION

AT THE TOP of the incline, at the level of the mine adit, the miners had a small building that was used as a workshop and warehouse. The miners' picks, the mine's tram cars, and other equipment were kept in working order here. A small forge was used to bend and temper the steel, and supplies for the mine were kept out of the rain. The entrance (adit) to the coal mine was on the south side of a small narrow draw. Rainstorms and snow made the center of the draw an often wet and unstable area, so the miners' building was built on the opposite (north) side of the draw. Here, Peter Bennett Sr., the son of Mine Foreman James Bennett, stood in front of the mine building, ready for another day's work. This photograph was taken as the camera lens was pointed north, near the area of the entrance of the mine.

Form 24 S-50,000-11-1-05

THE UINTAH RAILWAY COMPANY
COAL TICKET

Date .... 190 ....

Station ....

Amount .... **Ton**

Eng. No .... *Engr.* ....

Car No .... *Initial* .... *Pocket* ....

THE UINTAH RAILWAY COMPANY
COAL TICKET

Date .... 190 ....

Station ....

Amount .... **Ton**

Eng. No .... *Engr.* ....

Car No .... *Initial* .... *Pocket* ....

THE UINTAH RAILWAY COMPANY
COAL TICKET

Date .... 190 ....

Station ....

Amount .... **Ton**

Eng. No .... *Engr.* ....

Car No .... *Initial* .... *Pocket* ....

THE UINTAH RAILWAY COMPANY
COAL TICKET

Date .... 190 ....

Station ....

Amount .... **Ton**

Eng. No .... *Engr.* ....

Car No .... *Initial* .... *Pocket* ....

RODGER POLLEY COLLECTION

made things even worse as the flames were pulled into the mouth of the mine and onto the solid coal walls around it. As the fire quickly spread near the entrance, time was already running out for Rogers down in the mine.

Down at the houses at Carbonera, there was no reason to think anything was wrong, until Rogers did not appear for the Easter Sunday lunch. Stepping outside his house and looking up at the mine, Hamp Bennett saw the smoke and yelled the alarm that every coal miner hopes never to hear: *"Fire!"* A couple of miners ran up the incline to the entrance, while a call was placed to the mine foreman, James Bennett, who was enjoying Easter in Grand Junction. Mr. Bennett quickly called the mines at Palisade and Cameo, Colorado, and Sunnyside, Utah, to send help. By late that afternoon, men and material were arriving by train and truck to help with the rescue. As the first professional firefighters and miners from surrounding communities were on their way, the men who were at Carbonera made a desperate attempt to access the mine and try to find John Rogers. The tragedy was magnified when two of the would-be rescuers were overcome by the fumes just inside the main entrance. A rope was tied onto the leg of a young man, Paul Wagaman, and he crawled into the mine below the billowing clouds of smoke and soot. When he found the men, he gave a hard tug to the rope, held onto the clothing of the men, and slowly was pulled from the burning mine. By that evening the news of the fire had gotten to all points on the Uintah Railway in the surrounding country, and help was showing up by the hour. Volunteers from the Gilsonite mines in Utah and section men along the Uintah came forward to help.

CARBONERA WAS LAID OUT in a valley wider than most of the rest of the canyon between Carbonera and Atchee, and flood waters had a tendency to fan out in the bottom, leaving much debris and sand behind. The large dikes indicated on the maps near Bridge 20-A were used to try to keep the water inside the banks of the wash and away from the town and the railway bed. The houses and corrals west of the track, between Mileposts 20 and 21, were the only inhabited residences between Carbonera and Atchee. A small ranch operated from this location for many years, using the bottom of the valley for pasture land and drinking water from a spring in a canyon to the west was available. A small pipeline was run from the spring to a large concrete cistern that had been dug into a sandstone outcropping above the valley. This cistern was used for water storage, and another small cistern, next to the house, was used for the everyday needs of the family. By the 1930's, this small house was used only on a seasonal basis, as a sheep-grazing location; however, it eventually fell to ruins.

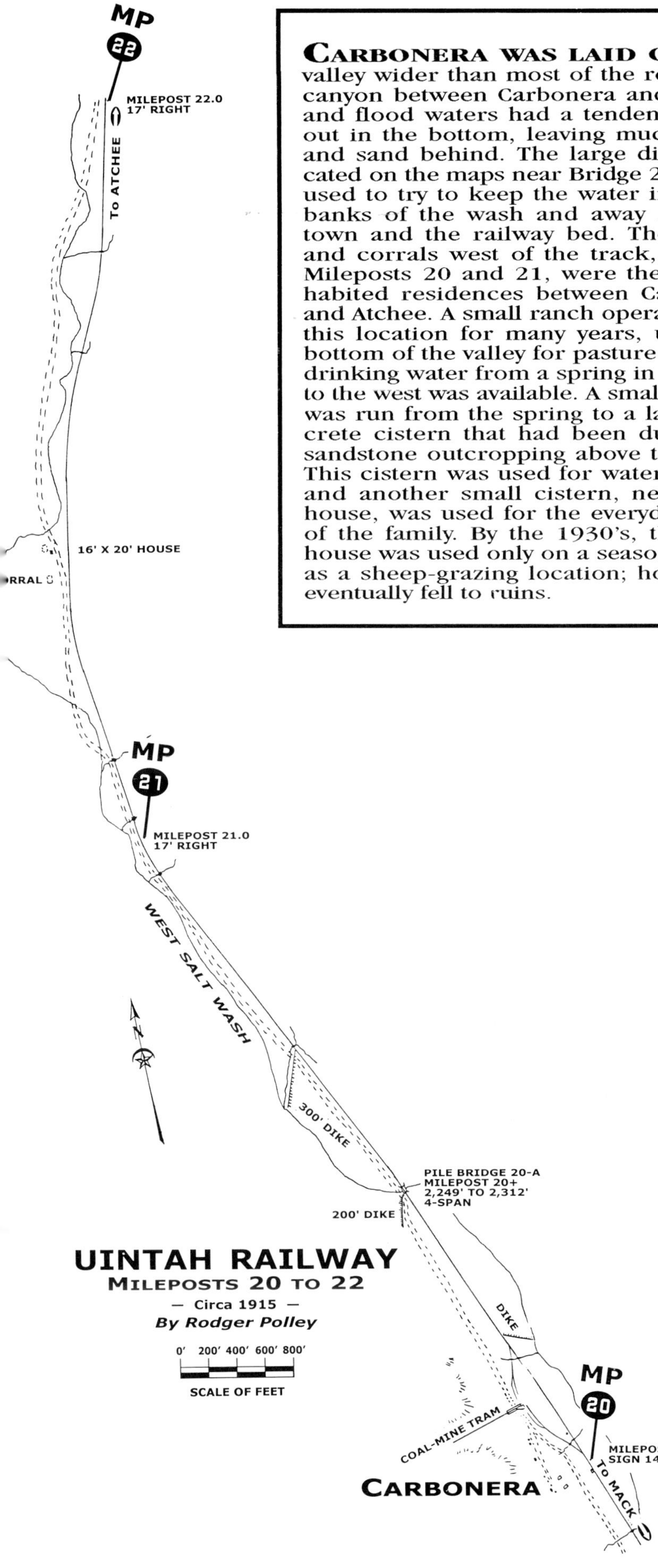

UINTAH RAILWAY
MILEPOSTS 20 TO 22
— Circa 1915 —
By Rodger Polley
0' 200' 400' 600' 800'
SCALE OF FEET

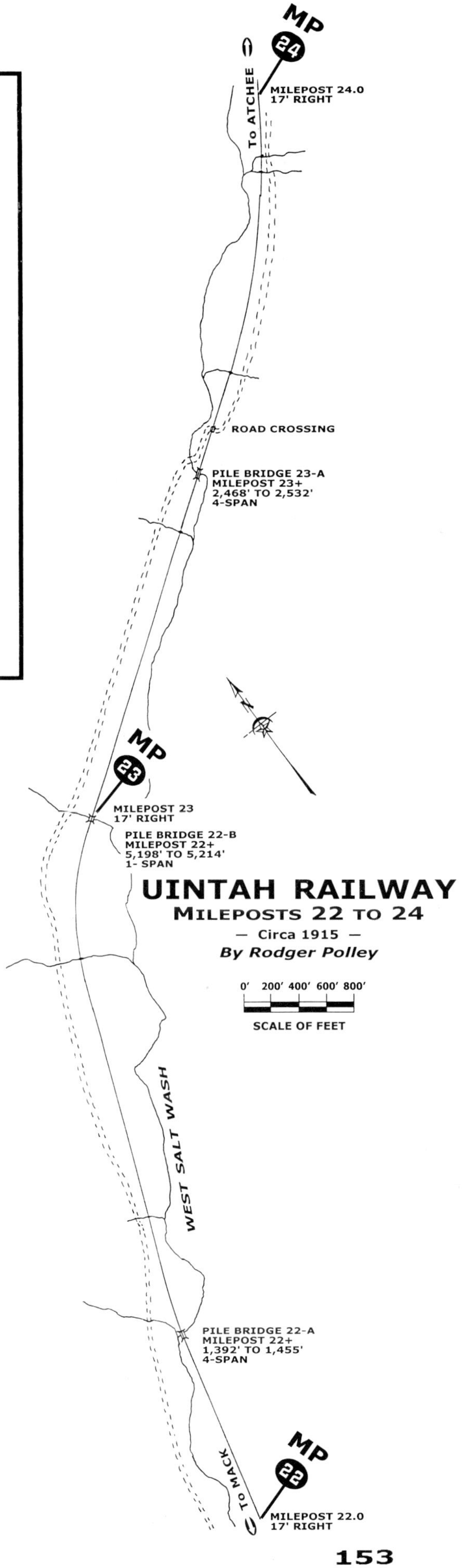

UINTAH RAILWAY
MILEPOSTS 22 TO 24
— Circa 1915 —
By Rodger Polley
0' 200' 400' 600' 800'
SCALE OF FEET

As the mine burned out of control above town, the boarding house and other residences were soon overcome by firefighters from all over. Various reports from the area went from 16 men fighting the fire, to 25 a few hours later, to 50 men by nightfall. That night, Colorado state mine inspector McClellan reached the scene and took charge of the operation. At this point, there was still limited hope that John Rogers was still alive in the depths of the mine, the fire only consuming the entrance way and not using the oxygen in the back of the mine. Mr. McClellan decided to lead another expedition into the mine, past the flames and into the rear of the mine. The party of men entered the mine through the small air shaft on the other side of the hill. The air inside was bad, and as they got closer to the main shaft, it became impossible to go any farther. The men pulled back just in time — and when they exited the mine, they all had to be treated for smoke inhalation, including the mine inspector.

The next morning saw even more supplies and manpower show up at Carbonera. The mine was still engulfed in flames, and the worker's boarding house was attempting to feed over 100 men. A "helmet crew" made up of men from the mines at Somerset and Cameo, breached the fire to the interior of the mine at 2:15 that afternoon. There, they found John Rogers and his dog, 450 feet from the front entrance. They figured that he had smelled the fire and had tried to make it out, running out of good air so close to the surface. With the recovery of the body, the focus turned to extinguishing the flames.

This was not going to be an easy task, as there was no water supply on this dry desert mountain. It was decided the only way to put the fire out was to suffocate it. By depriving the fire of oxygen, hopefully it would quickly burn itself out. The two tunnels into the mountain would have to be sealed. The main entrance was easy; it had the large steel door that could be shut and an airtight seal made from the outside. The fan shaft was a different matter. It was decided to bring part of the mountain down over the top of the shaft.

THE UINTAH RAILWAY COMPANY

1919 TERM PASS No. Q188

PASS ---V. L. McCoy---

ACCOUNT Engineer

BETWEEN All Points

UNTIL Dec. 31st 1919 UNLESS OTHERWISE ORDERED AND SUBJECT TO CONDITIONS ON BACK

VALID WHEN COUNTERSIGNED R. D. Clark.

COUNTERSIGNED R D Clark

GENERAL MANAGER

RAND MCNALLY & CO.

RICHARD A. RONZIO COLLECTION

ORIGINAL ART BY HOWARD FOGG
AL & BERNADETTE CHIONE COLLECTION

UINTAH RAILWAY CO.
51
51

OTTO C. PERRY PHOTO – RICHARD H. KINDIG COLLECTION

THE RESPECTED Otto C. Perry was doing a great job of train chasing on May 9, 1939. He took several good photographs of No. 50 hauling a load up the canyon to Atchee. Here, Otto caught the train halfway across Trestle 22-B, a few miles past Carbonera. This section of the canyon was never used by farmers or ranchers, as the bottom of the canyon floor continued to get narrower, and West Salt Wash became much deeper in the bottom of the wash.

Dynamite was installed and the tunnel was blasted shut on top of the fan. Details as to how long this process took or how things worked during the clean-up were lost. As for the Uintah, things were pretty much at a standstill until the mine could be reopened. The workers were hired out to get the job done, and within the next few weeks the main entrance was cleaned up and re-timbered. A new fan shaft was put in on the left side of the main entrance, utilizing the same system for sealing off the main entrance with the iron door. That was how the mine looked on the outside until it closed down in 1939.

As the rails were pulled up along the Uintah's right-of-way during the summer and fall of 1939, the mine trestle and siding were torn out before the main line to Atchee was even started on. Several heaping loads of coal destined for the coal bunkers at Atchee and Mack were the last loads to leave. Like the rest of the small ghost towns that the Uintah left behind, Carbonera was left to the scrappers. Timbers, wood

planks and any other usable material were systematically stripped by local subcontractors, who used the materials for private purposes or resold them to other people. What was left — odd lumber, bent tin and pieces not worthy of the contractor's labor — was quickly picked up by local farmers and people in Fruita and Grand Junction looking for free wood, hot on the heels of the Great Depression. *So... faster than it was built, Carbonera disappeared from the scene!*

Today, only a few reminders of Carbonera are left. The cut in the hill where the incline was built is discernible, if one knows where to look. The large cement slabs that made up the base for the coal tipple are still standing in the sagebrush, as are a few dugout cellars in the ground. For the most part, Nature has reclaimed the entire area and, except for the current dirt road on the old railway bed and a few pipeline location markers, there is nothing there that does not look the same as it did 100 years ago, before the Uintah Railway arrived.

Form 2121 7-38 100

THE UINTAH RAILWAY COMPANY

Employees' Time Table
No. 26

To take effect Sunday, July 31, 1938, at 12:01 a. m.

Mountain Standard Time

Superseding Time Table No. 25 and Supplements Thereto.

For the government and information of employees only, and not intended for the use of the public

The company reserves the right to vary from this Time Table at Pleasure.

J. I. McCLAIN, Acting General Manager.

E. V. EARP, Superintendent.

ENGINE RATING

WESTBOUND

From Mack, Colo., to Carbonera, Colo.:

Class Eng. 40	300 tons
Class Eng. 50	525 tons
Class Eng. 51	525 tons

From Carbonera, Colo., to Atchee, Colo.:

Class Eng. 40	225 tons
Class Eng. 50	525 tons
Class Eng. 51	525 tons

From Atchee, Colo., to Baxter Pass, Colo.:

Shay Class Eng. 5 and 7	60 tons
Class Eng. 50	90 tons
Class Eng. 51	90 tons

From Rainbow Junction, Utah, to Rainbow, Utah:

Shay Class Eng. 5 and 7	105 tons
Class Eng. 30	110 tons
Class Eng. 50	200 tons
Class Eng. 51	200 tons

EASTBOUND

From Watson, Utah, to Dragon, Utah:

Class Eng. 30	650 tons
Class Eng. 50	1150 tons
Class Eng. 51	1150 tons

From Dragon, Utah, to Wendella, Colo.:

Class Eng. 30	325 tons
Class Eng. 50	525 tons
Class Eng. 51	525 tons

From Wendella, Colo., to Baxter Pass, Colo.:

Shay Class Eng. 5 and 7	105 tons
Class Eng. 50	185 tons
Class Eng. 51	185 tons

STAFF

J. G. Griffin — Master Mechanic

E. C. Dunlap — Car Foreman

J. L. BOOTH, Chief Dispatcher

WESTBOUND — EASTBOUND

Telephone Calls	Telegraph Calls	No. 1 First Class Leave Biweekly Mon. and Fri.	Miles From Mack	STATIONS AND SIDINGS	Miles From Watson	No. 2 First Class Arrive Biweekly Tues. and Sat.	Car Capacity of Sidings, Location of Water, Fuel and Turning Stations
--	A	11:30 a	0	MACK D 4.2	62.8	1:30 p	45 Y W C
		F 11:40	4.2	CLARKTON 7.3	58.6	F 12:35	7
		F 12:01 p	11.5	SPRAGUE 6.0	51.3	F 12:20	14
		12:15	17.5	COOLEY 2.6	45.3	12:01 p	10
----		F 12:30	20.1	CARBONERA 8.2	42.7	F 11:55	24 C
---		S 1:15 1:45	28.3	ATCHEE 1.8	34.5	S 11:35 11:05	74 Y W C
		1:55	30.1	MORO CASTLE 2.1	32.7	10:40	
		2:15	32.2	SHALE 2.0	30.6	10:30	10
---		S 2:40	34.2	BAXTER PASS 1.9	28.6	S 10:15	17 Y
		2:50	36.1	DEER RUN 1.3	26.7	9:55	6
		F 3:00	37.4	COLUMBINE 2.5	25.4	F 9:45	8 W
---		F 3:15	39.9	McANDREWS 1.0	22.9	F 9:22	9
		S 3:20 3:40	40.9	WENDELLA 1.2	21.9	S 9:15	22 Y W C
		3:42	42.1	SEWALL 4.5	20.7	8:55	7
		3:55	46.6	EAST VAC 1.4	16.2	8:40	9
		3:57	48.0	URADO 5.3	14.8	8:30	2
----		4:20 4:40	53.3	DRAGON 1.3	9.5	8:20 8:00 a	36 Y W C
		4:42	54.6	COUNTRY BOY 1.1	8.2	7:52	7
		4:47	55.7	RECTOR 2.9	7.1	7:50	7
		4:50	58.6	UTE 3.5	4.2	7:40	6
		5:05	62.1	RAINBOW JUNCTION 0.7	0.7	7:32	Y
—	WN	5:30 p	62.8	WATSON D	0	7:30 a	29
		Arrive Biweekly Mon. and Fri.				Leave Biweekly Tues. and Sat.	

SPECIAL RULES AND INSTRUCTIONS

No. 1—Eastbound trains are superior to Westbound trains of the same class except:

1-A—Second class and extra trains must clear the time of No. 1 and No. 2 ten minutes.

No. 2—No train will leave Mack or Watson without clearance card.

No. 3—Train Register Books are located at Mack, Atchee and Watson.

3-A—Bulletin Books Mack, Atchee and Watson.

No. 4—Yardlimit stations are Mack, Carbonera, Atchee, Wendella, Dragon, Rainbow Junction, Rainbow and Watson.

No. 5—Trainmen must inspect and see that all retainers are turned up before starting down grade and they must assist when necessary with hand brake to insure safety of the train.

5-A—All trains must be thoroughly inspected and air-tested by car men and trainmen jointly at Atchee before leaving.

5-B—On all trains at Baxter Pass, train and enginemen must inspect train and test air; while the air is applied trainmen must turn up retainers on each car to ascertain their working condition; examine all hand brakes and confer with engineer and know that everything is in working order and in safe condition before starting their train down the heavy grade.

5-C—Enginemen and trainmen must use every precaution against the parting of trains on heavy grades. In case of trouble, train must be stopped, a careful inspection made and defects remedied if possible and report made to Superintendent.

5-D—Trainmen must not uncouple cars on grade without first testing hand brakes and knowing that they will hold. Cars placarded EXPLOSIVES or INFLAMMABLE must not be cut off while in motion.

5-E—Trainmen must be out and located where they can give signals and apply air when passing over bad track or bridges covered by slow order or otherwise, and give proceed signal to enginemen after passing, which signal must be answered by the engineer.

5-F—One member of train crew must be on the rear car of train when both ascending and descending grades. All cars in train must be closely watched by trainmen and see that wheels are not sliding and damaged.

5-G—All enginemen should run carefully and approach bridges liable to wash slowly during violent rain storms or immediately thereafter to avoid accidents.

No. 6—All employees are hereby notified that there are coal chutes, buildings, platforms and other structures located on main track and on sidings, also structures and platforms belonging to private corporations and persons located on industrial sidings and spurs that WILL NOT CLEAR a man riding on the side or top of car or engine, and all employees must protect themselves from injury in passing such structures.

No. 7—The speed of trains should be so restricted that absolute safety will be assured and the maximum speed will ordinarily be that which is necessary to make the schedule, except in territory specified enginemen and conductors may moderately increase the speed above that required by the schedule when the necessity arises, if in their judgment it would be safe and prudent, bearing in mind that safety in operation is to be given first consideration and always giving due consideration to track and surrounding conditions.

7-A—All trains must use fifty minutes descending heavy g tween Baxter Pass and Atchee, and forty minutes betwee Pass and Wendella.

No. 8—The following signs indicate:

S—Regular stop.

F—Flag stop to receive and discharge passengers or freight.

D—Day only telegraph office.

W—Water.

C—Coal.

Y—Wye.

No. 9—Persons accompanying live stock or other freight will ried on any freight train handling such live stock or freigh holding proper transportation and when permission to acc same is covered by contract. Passengers on freight trains sh informed that cabooses will not be pulled up to platform to or deliver passengers or baggage.

9-A—Employees holding passes will be carried on any freig to and from points at which trains stop, when passes are "Good on Freight Trains."

No. 10—Prompt report must be made to Superintendent of dents and unusual occurrences.

10-A—When an accident, collision of trains or any coll trains with vehicle or pedestrian, resulting in loss of lif jury to persons in Colorado, the superior officer, the employee on the ground at time of such accident shall ately notify the Public Utilities Commission, Capitol B Denver, Colorado, by telegram, the details of such accide ing the immediate location and nature of the accident number of persons killed or injured. Information coveri accidents must be sent by Western Union Telegraph C wire, and all agents will accept and so transmit same, a charges therefor against the Railroad account.

No. 11—The following are auxiliary lines:
The Rainbow branch.

11-A—Trains on the Rainbow Branch will be operated un yardlimit rule.

No. 12—Enginemen must sound whistle approaching all cros

EXTRACTS FROM
REVISED STATUTES OF UTAH

8203. (4203.) Placing Freight Car in Rear of Train.—Eve son who, in making up or running railroad trains, places or causes to be placed or run, any freight car in the rear of pa cars is guilty of a misdemeanor, and, if loss of life or limb from such placing or running, is guilty of a felony. The term " car" as used in this section does not include a baggage, exp mail car.

(For use in connection with Uniform Domestic Straight Bill of Lading, adopted by Carriers in Official, Southern, Western and Illinois Classification territories, March 15, 1922, as amended August 1, 1930.)

2 **THIS SHIPPING ORDER** must be legibly filled in, in Ink, in Indelible Pencil, or in Carbon, and retained by the Agent.

Shipper's No.______

Agent's No.______

THE UINTAH RAILWAY COMPANY

RECEIVE, subject to the classifications and tariffs in effect on the date of the issue of this Shipping Order.

At______ 193__

From______

the property described below, in apparent good order, except as noted (contents and condition of contents of packages unknown), marked, consigned, and destined as indicated below, which said company (the word company being understood throughout this contract as meaning any person or corporation in possession of the property under the contract) agrees to carry to its usual place of delivery at said destination, if on its own road or its own water line, otherwise to deliver to another carrier on the route to said destination. It is mutually agreed, as to each carrier of all or any of said property over all or any portion of said route to destination, and as to each party at any time interested in all or any of said property, that every service to be performed hereunder shall be subject to all the conditions not prohibited by law, whether printed or written, herein contained, including the conditions on back hereof, which are hereby agreed to by the shipper and accepted for himself and his assigns.

(Mail or street address of consignee—For purposes of notification only.)

Consigned to______

Destination______ State of______ County of______

Route______

Delivering Carrier______ Car Initial______ Car No.______

No. Packages	DESCRIPTION OF ARTICLES, SPECIAL MARKS, AND EXCEPTIONS	*WEIGHT (Subject to Correction)	Class or Rate	Check Column

Subject to Section 7 of conditions, if this shipment is to be delivered to the consignee without recourse on the consignor, the consignor shall sign the following statement:

The carrier shall not make delivery of this shipment without payment of freight and all other lawful charges.

......

(Signature of consignor.)

If charges are to be prepaid, write or stamp here, "To be Prepaid."

......

Received $......
to apply in prepayment of the charges on the property described hereon.

......

Agent or Cashier.

Per......
(The signature here acknowledges only the amount prepaid.)

Charges Advanced:

$......

* If the shipment moves between two ports by a carrier by water, the law requires that the bill of lading shall state whether it is "carrier's or shipper's weight."

Note—Where the rate is dependent on value, shippers are required to state specifically in writing the agreed or declared value of the property.
The agreed or declared value of the property is hereby specifically stated by the shipper to be not exceeding ...... per ......

______Shipper.

Per______

☞ **Agent must detach and retain this Shipping Order and must sign the Original Bill of Lading.**

Permanent postoffice address of shipper______

J. J. COLLINS' SONS INC., PRINTERS, CHICAGO

RODGER POLLEY COLLECTION

RODGER POLLEY COLLECTION

THE ROCKY TERRAIN through which the Uintah ran from Carbonera to Atchee is exemplified in this photograph of Bridge 23-A. Huge layers of sandstone jut up from the canyon floor, creating a narrow canyon, through which water from the above mountains had to pass. In the foreground is the automobile road, winding its way through the sagebrush and gullies, while the Uintah Railway's grade maintained its steady incline toward Atchee, at the right. Bridge 23-A was typical of most of the bridges on the Uintah line, short and built without a walkway. Although a large culvert would have been adequate to carry the usual trickle of water that came down West Salt Wash Creek, the constant threat of flash floods required the use of wooden pile bridges.

ON A TRIP BACK from college, Roy Eno Jr. was standing next to his father and longtime Uintah engineer, Roy Eno, for this picture. They were in front of the remains of the stock cars. These men from two generations became well-known railroad men in western Colorado. Roy Jr. ended-up working for the Denver & Rio Grande Western until retirement. Roy Sr. was a well-respected engineer wherever he went. Roy Sr. and Vic Earp had the most seniority as engineers on the Uintah when it closed down. They both had been assigned to the two articulated engines. Many people were not happy about the Uintah Railway's abandonment, and they showed their displeasure to the employees who stayed to work on the wrecking trains. Roy Sr. was one of the men who stayed, but he did not do so for the money, as much as the fact that he loved working on the Uintah more than anywhere else, and he would operate a locomotive as long as anyone would let him.

ROY ENO JR. COLLECTION

ROY ENO JR. COLLECTION

LATE IN SEPTEMBER OF 1939, a train containing several empty Uintah flatcars, stock cars, a coal-laden gondola and one of the little four-wheel cabooses had "tied down" in the yard at Atchee, en route northbound. This freight train had been made up for the scrapping crews, who were pulling up the rail near Watson and Dragon, in Utah. Two train crews, made up of Uintah employees, were utilized to operate the trains over Baxter Pass, but many of the contractors doing the scrapping were employed to help. It is not known how the train started to roll, but speculation is that someone from the scrapping crew had not set the handbrakes properly. By the time, someone noticed the train was moving, it was impossible to do anything but watch it roll out of town, heading downgrade toward Mack. Rumbling downgrade, caboose first, the runaway lost its first car just outside Atchee, off the end of the consist. Three miles south of Atchee—just south of Bridge 24-D—the speed of the cars became too much for the diminutive caboose, as its wheels screeched around a curve. The caboose promptly left the rails, followed by several stock cars—and the whole pileup was pushed over onto itself by the weight and speed of the following cars. The stock cars piled on top of one another after leaving the rails, and they went in all directions. Both of these photographs were taken looking upgrade toward Atchee. Although the stock cars were smashed, many of them retained their shape, while the string of flatcars behind them was in worse shape. The empty flatcars quickly piled on top of one another like pancakes—leaving their trucks at the bottom and elevating one another up into the air.

ROY ENO JR. COLLECTION

THE GONDOLA at the end of the runaway consist slowed as the cars piled up on top of each other. The cars slowly came to a stop before tipping over to one side.

ROY ENO JR. COLLECTION

THE REMAINS of the four-wheel caboose, which had been at the front of the runaway cars, are the subject of this view. The little "hack" was the first of the cars to leave the track, and it rolled over and over, in front of the rest of the cars, scattering debris in all directions. Here, the deck and end sills were all that remained of this diminutive caboose. The superstructure of the car had been ripped away, leaving only the brake wheels and a ladder attached. A pair of trucks from one of the stock cars was sitting on the embankment, waiting for the scrappers to sort through the pile of wreckage to salvage parts. Many onlookers from Atchee, Mack and Fruita traveled up the canyon by automobile to view the wreck, after word about it spread. In this scene, Ruby Luton, post mistress in Atchee, was standing next to the door of the caboose, which had been thrown over 50 feet through the air during the crash—without the glass being broken!

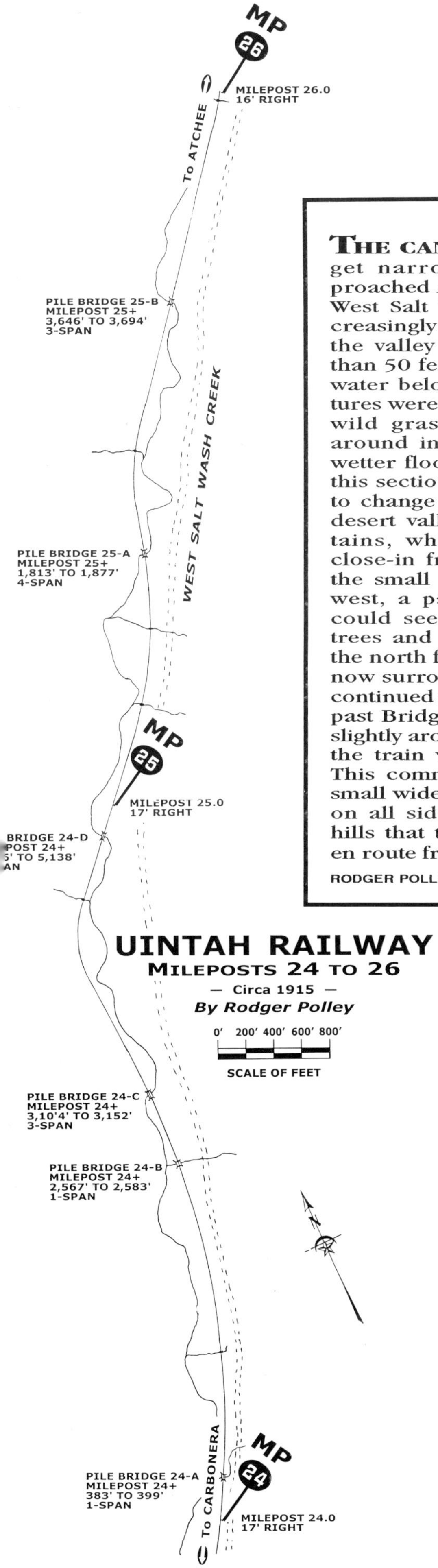

The canyon continued to get narrower as the track approached Atchee. At the same time, West Salt Wash Creek became increasingly deeper in the bottom of the valley – in some places more than 50 feet from track level to the water below. A few fenced-off pastures were along the right-of-way, as wild grass grew nearly all year around in the slightly cooler and wetter floor of the valley. It was in this section that the climate started to change ever-so-slightly from the desert valley to that of the mountains, which were beginning to close-in from both directions. Up the small canyons to the east and west, a passenger on the Uintah could see the groves of tall pine trees and aspens, spotted all over the north faces of the high hills that now surrounded them. The canyon continued to get narrower, until just past Bridge 27-A, the track curved slightly around a hill, and in front of the train was the town of Atchee. This community was situated in a small wide-open valley, surrounded on all sides with the same desert hills that the train passed through en route from Carbonera.

RODGER POLLEY COLLECTION

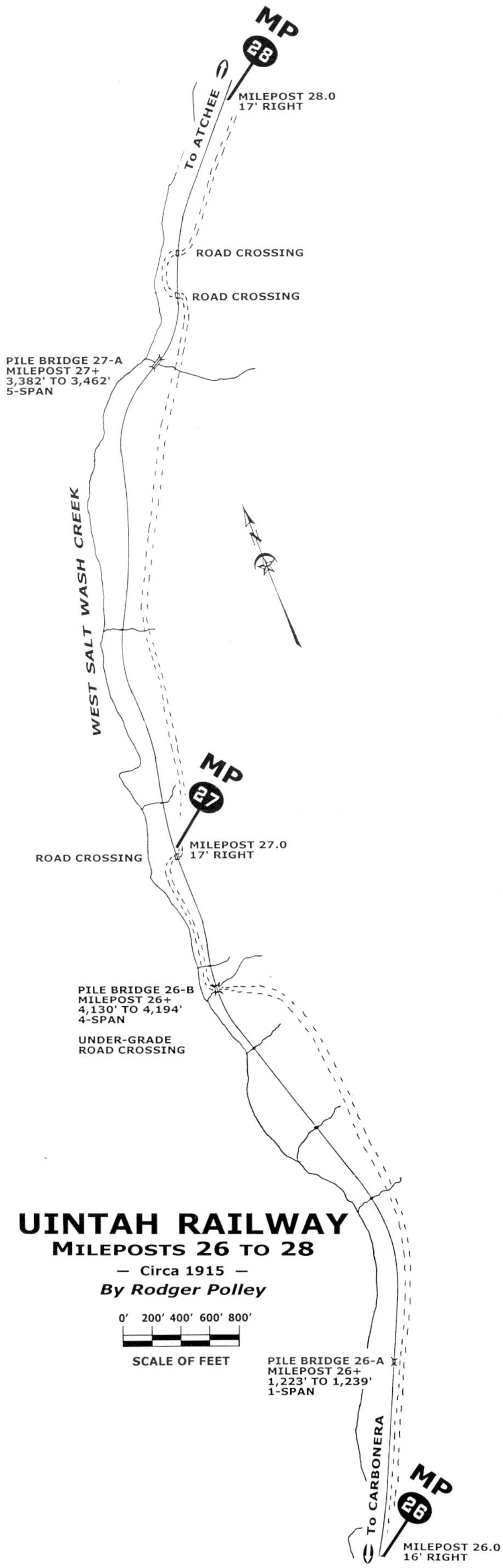

OTTO C. PERRY PHOTO – RICHARD H. KINDIG COLLECTION

THE SAME TRAIN that Otto C. Perry had been following on May 9, 1939, made its way through a small cut nearing Milepost 27.0. The power-

ful articulated locomotive No. 50 was pulling hard on the load of stock cars and boxcars, combination passenger car No. 50 and four empty water cars.

RICHARD B. JACKSON PHOTO – BILL ELLICOT COLLECTION

ON HER FINAL RUN up to Atchee from Mack, the Uintah's 2-8-0 No. 12 was rolling along before being shipped to her new owner, the Eureka–Nevada Railway in Palisade, Nevada, during 1937. The engine had stopped on Bridge 27-A to pose for a picture for Richard B. Jackson from the automobile road east of the tracks. The extra-long trestle and heavy timbers of Bridge 27-A were needed to span the small wash that came in from the left. Although never running with much water, the steep walls of the wash were made entirely of decomposed earth and loose dirt, and it was impractical to maintain a center bent, as any great flow of water from above washed the bridge out.

THE TAIL-END of a northbound train was captured on film as it rounded a curve en route to Atchee. Steel-sheathed combination car No. 50 was trailing along at the back of a mixed train, which consisted of empty flatcars. Black smoke was pouring out of the smokestack, as the engine was pulling up the steady 2.0-percent climb, while in the distance the steep grade up Baxter Pass was ahead—in defiance of the approaching train. The flatcars were destined for the Gilsonite mines in Utah. Notice the rough road beside the track.

ROBERT C. BOULTER COLLECTION

RICHARD B. JACKSON PHOTO – RICHARD H. KINDIG COLLECTION

ENGINE No. 12 was hauling a train upgrade near Milepost 27.0, on August 9, 1937. Empty water cars were headed for Atchee, flatcars were going to the mines, while a gondola full of coal, a boxcar full of freight and the customary combine could be seen at the end of the mixed train.

JIM MCCOY COLLECTION

DURING AN EARLIER year, Engine 12 was near Milepost 27, pushing the wedge plow through freshly fallen snow, up toward Atchee. It is interesting to notice that the snowplow was normally kept at Atchee, and it must have been taken south the night before.

ROBERT C. BOULTER COLLECTION

LOCOMOTIVE No. 10 was pulling into Atchee with a string of stock cars, and a helper engine had been cut into the train for the upgrade run to Atchee. This livestock extra was run for ranchers during the spring, so they could graze their animals in the higher green elevations of the area surrounding Baxter Pass.

OTTO C. PERRY PHOTO
DENVER PUBLIC LIBRARY,
WESTERN HISTORY DEPARTMENT

ALTHOUGH SHAY No. 7 did take an occasional extra train or snowplow to the top of Baxter Pass, most of her life during the 1930's was spent doing mundane switching, as needed in the yard at Atchee. Here, this Shay was getting ready to pull three empty water cars off the train, and go about switching the rest of this long 15-car train into manageable sections for the haul over Baxter Pass on May 9, 1939. The track in the foreground was part of the wye, and most switching was done from here to the east.

ROBERT C. BOULTER COLLECTION

ENGINE No. 40 was working hard, heading upgrade into town on a cold winter day in 1938. She was pulling a long train consisting of coal from Carbonera, two empty water cars and a long string of empty flatcars. The train would pull forward to the water tank, where the cars would be split up. The engine, gondolas and water cars would be switched out, along with the caboose or combination car at the end of the train. Then, the flatcars would be split up for the pull up the 7.5-percent grade on this side of Baxter Pass. Notice the wooden platform and stairs at the end of the boardwalk, for access to the track level. Snow covered portions of the ground in this picture.

CHRIS H. UEBERROTH COLLECTION

A THIN LAYER OF SNOW had coated the ground before this winter picture was taken. Engine No. 12 was pulling a long freight train into Atchee from Mack. The mainline ran directly in front of the company houses—literally at their front gate. As was the standard procedure on

most railroads, empty low-weight cars such as flatcars were cut in toward the rear of the train. This was especially true on the Uintah's climb over Baxter Pass, as too much weight toward the rear of the train while braking could lead to bunching and derailments during descents from the pass.

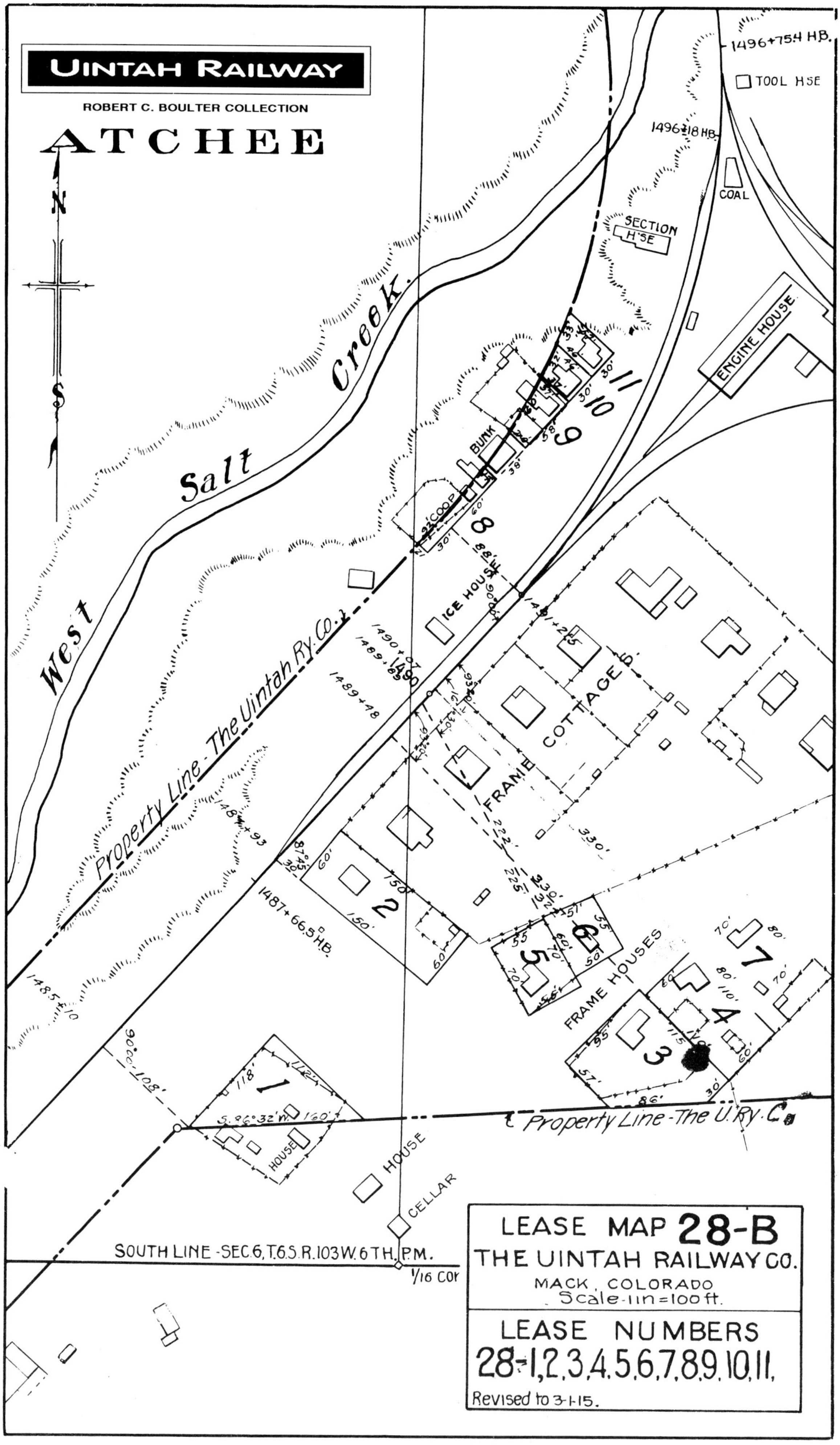
UINTAH RAILWAY
ROBERT C. BOULTER COLLECTION
ATCHEE
N
S
West Salt Creek
Property Line - The Uintah Ry. Co.
TOOL H'SE
COAL
SECTION H'SE
ENGINE HOUSE
BUNK
COOP
ICE HOUSE
FRAME COTTAGE
FRAME HOUSES
HOUSE
CELLAR
Property Line - The U. Ry. Co.
SOUTH LINE - SEC. 6, T. 6 S. R. 103 W. 6 TH. P.M.
1/16 COR
1496+75.4 HB.
1496+18 HB.
1490+07
1489+85
1489+48
1487+66.5 HB.
1485+10
LEASE MAP 28-B
THE UINTAH RAILWAY CO.
MACK, COLORADO
Scale - 1 in = 100 ft.
LEASE NUMBERS
28-1, 2, 3, 4, 5, 6, 7, 8, 9, 10, 11.
Revised to 3-1-15.

Chapter 3

An Isolated Community...
ATCHEE, COLORADO

IT IS UNCLEAR as to when the first school building was constructed in Atchee, but it probably was in 1908 or 1909. The first building was constructed by the railway for the employees who lived in Atchee. That building was built east of the residential part of town in the wide canyon behind Atchee. This was the typical one-room schoolhouse, with front and back doors, three windows on the south side and two on the north side. A two-hole outhouse was constructed behind the school, and a large boardwalk was extended from the boardwalk in front of the company houses up to the school and "out to the outhouse."

Within a few years, the first school building became inadequate for the number of children in the area. Not only were there more men with families working for the Uintah Railway, but homesteaders were making their way into Salt Wash Canyon, and a few small ranches were popping up here and there. The front of the building was added onto sometime before 1910. The A- frame construction was extended toward the front of the building, gaining enough room to create a two-room school, one for the first four grades and one for grades five through eight.

The construction created not only a larger more practical school, but also a much more "school-looking" building, compared with the small cottage it replaced. The schoolhouse now was lined on both sides with four windows, allowing enough light to get into the classrooms. At the front, a bell tower and a small picture window were installed above the front door. Since the terrain the building sat on consisted of a steady upgrade toward the back of the building, the extension made the front of the building and front door quite a bit higher than it was before. New stairs were built along both sides of the door up to the level of the door. In front of the stairs, a large wooden platform filled-in the space between the boardwalk and the building. Other improvements to the building included painting the outhouse, building a swing set and clearing away most of the sagebrush in a large area around the school. Although sagebrush can be beautiful to someone from the city, most Westerners soon found that snakes, gnats and ticks were found less frequently when most of the sagebrush was removed.

This building was used by the town of Atchee through World War I. As production of Gilsonite at the mines in Utah increased, and the Uintah Railway continued to use a large work force to keep the engines running, Atchee was slowly growing. Need for a larger, more modern school building was a constant concern among the adults in Atchee. By 1920, the school at Atchee was being recognized by Garfield County and was known as School District No. 39. By the early 1920's, the citizens at At-

(Continued on Page 198)

COLORADO RAILROAD MUSEUM COLLECTION

ATCHEE WAS NAMED for Chief Sam Atchee, a Uintah chief, who was part of the political leadership of the Ute tribe during the building of the railway in 1904. A brother-in-law of Chief Ouray and related to his wife, Chipeta, Atchee was a well-known and respected Indian chief in Colorado and Utah.

RODGER POLLEY COLLECTION

THIS EARLY VIEW portrays the front of the enginehouse and side of the concrete machine shop. A section gang was working on the track that curved around the enginehouse and completed the wye. The "mainline" through the yard was the track curving to the left. The track the passenger car was sitting on was a passing siding, which stretched from the east end of town, almost to Bridge 28-A. The pile of earth at the left of the passenger car was created when the yard area was lowered during construction. This loose shale dirt was a constant nuisance to the track crews in the yard at Atchee, because the flow of mud that filled-in between the ties and rail, causing slow-speed derailments. Obviously, it was time for maintenance again, as the mud was nearly level with the topof the rails. The small building with the peaked roof, in front of the enginehouse, was the original yard office. A few years after it was built, the railway abandoned this structure. It then was moved to the front of the coal chute and used as a sandhouse Both enginehouse doors were open, and sitting in the right entrance was one of the Uintah's small railcars. There is little adequate information on these cars—even whether or not there was more than one—but there are many fuzzy and distant photographs showing a type of motor car, unlike the Mack railcar so commonly photographed. In the west stall you can see all the way to the far end of the long building, as a window on the far wall allowed light into the rather dark building.

THIS ENLARGED SECTION of the previous picture was taken prior to the school house being built. Bridge 28-A cast its shadow on West Salt Wash in the late afternoon. Shay No. 1 was sitting on the wye in this otherwise uneventful view of the coal bins, water tank and enginehouse. Company houses and the hotel are in the background.

JOHN AND HAZEL CARSON COLLECTION

JOHN AND HAZEL CARSON COLLECTION

THIS PICTURE ANGLE of Atchee was taken from the vicinity of Rogers' ranch, and it shows the stark little town-setting, surrounded by sagebrush.

DEL GERBAZ COLLECTION

DURING THE FIRST 15 YEARS of the Uintah's life, passenger traffic often consisted of two passenger cars, to accommodate travelers going to the little towns along the line. Being the closest railway connection to Vernal, Utah, and the other small farming towns in the Uintah Basin, the Uintah was kept busy with the comings and goings of the local population. Before automobiles, the Uintah was *the* way to get in and out of the basin. This view of the second No. 11 was taken on June 30, 1914, in the yard at Atchee, just after she had arrived from Mack. Then, the engine would uncouple from the train, pull forward up to the mainline to Bridge 28-A, and then back onto the wye to prepare for the return trip to Mack. The wooden steps going up to the level of the boardwalk went directly to the front of the company hotel.

ON JULY 30, 1914, excursionists were on their way to Vernal, Utah, and the Uintah Basin in a two-car passenger train. The cars had just been brought up from Mack behind the Uintah's second engine, No. 11. That engine then uncoupled from the train and one of the tank engines was placed in front of the train for the long climb over Baxter Pass. A Shay engine was then coupled on to the front of the tank engine to be used as a helper up the steep 7.5-percent grade on the south side of the pass. At the top of the pass, the Shay would cut off from the train and come back to Atchee. This picture was taken from the end of the boardwalk looking west towards the engine house and station platform. Notice the material-storage shed that was in use at this time, on the east side of the enginehouse.

RICHARD A. RONZIO COLLECTION

COLORADO RAILROAD MUSEUM COLLECTION

BEAUTIFULLY LETTERED Baldwin-built Consolidation No. 10 was built in 1904. She played the role of being the favorite as a picture subject, much the same way No. 40 would do on the same route decades later. Always kept in immaculate shape, this 2-8-0 was the motive power routinely used on passenger trains between Mack and Atchee. She retained that role for nearly all of the first 23 years of operation on the Uintah Railway.

FRANK A. KENNEDY PHOTO – RICHARD A. RONZIO COLLECTION

THIS PANORAMA OF ATCHEE was recorded by "Judge" Frank A. Kennedy around 1907. All three of the Uintah's ex-D&RG Pullman cars were sitting on the siding, along with one of the wooden water cars and a tender from some engine. The Uintah utilized several old tenders that were bought secondhand as water-service cars during the early years. These cars later spurred the building of the steel water cars instead of retaining the high-maintenance wooden water cars. Notice how the boardwalk in front of the company houses was, right up against the porches and how it cut across the yard of the hotel at an angle. The tents in the foreground were fairly permanent structures, utilizing a floor and wooden frame to hang the canvas on. Notice the two to the left are missing their canvas, leaving only the half walls on the bottom and the wooden frame standing.

FRANK A. KENNEDY PHOTO – DAN ABBOTT COLLECTION

SINCE THE UINTAH RAILWAY had relatively few engines to be serviced, there was no need for the expense of building a turntable at Atchee. As a result, Atchee was laid out with a simple wye. The end of the wye at this time was just past the main coal tipple. Later, it would be extended and a three-way switch was installed closer to the hill allowing a train to switch out a car to the car shops or to the coal tipple. Here the coal tipple has not been built to the top of the coal bunker and all of the coal was shoveled by hand into the bunker. Most coaling for engines that were dispatched from Atchee in the morning already had a full load of coal filled by the hostlers, while locomotives going through Atchee or on their way back to Atchee coaled from the smaller bunker next to the water tank on the mainline. This small bunker was removed between 1915 and 1919, to allow room for a second water tank. Notice the small stair steps up the left side of the smaller bunker. The smokestack from the original stationary boiler, which was on the east side of the machine shop, can be seen sticking up behind the water tank. The track layout of the wye was very simple at this time, with only a small stub track coming down from the back of the wye to spot a gondola full of coal for the boiler. That track later would be extended to the area of the small coal bunker, and it tied back into the rest of the track.

STANDING ON TOP of two of the Uintah's flatcars in the yard at Atchee were three men, from left to right: Misters Stillwell, Dixon and Cox. The large load of hay probably was destined for Dragon or Watson for use by the Uintah Toll Road Company's many teams of horses. Observe the difference in the new paint on flatcar No. 116 and the disheveled appearance of the car in front of it. In the background, a gondola and flatcar of sacked Gilsonite ore were waiting for a train to take them to Mack. The picture was taken on May 20, in an unknown year, as a light skiff of snow had covered everything, showing just how late spring could come in the Colorado mountains.

COLORADO RAILROAD MUSEUM COLLECTION

UINTAH RAILWAY
116

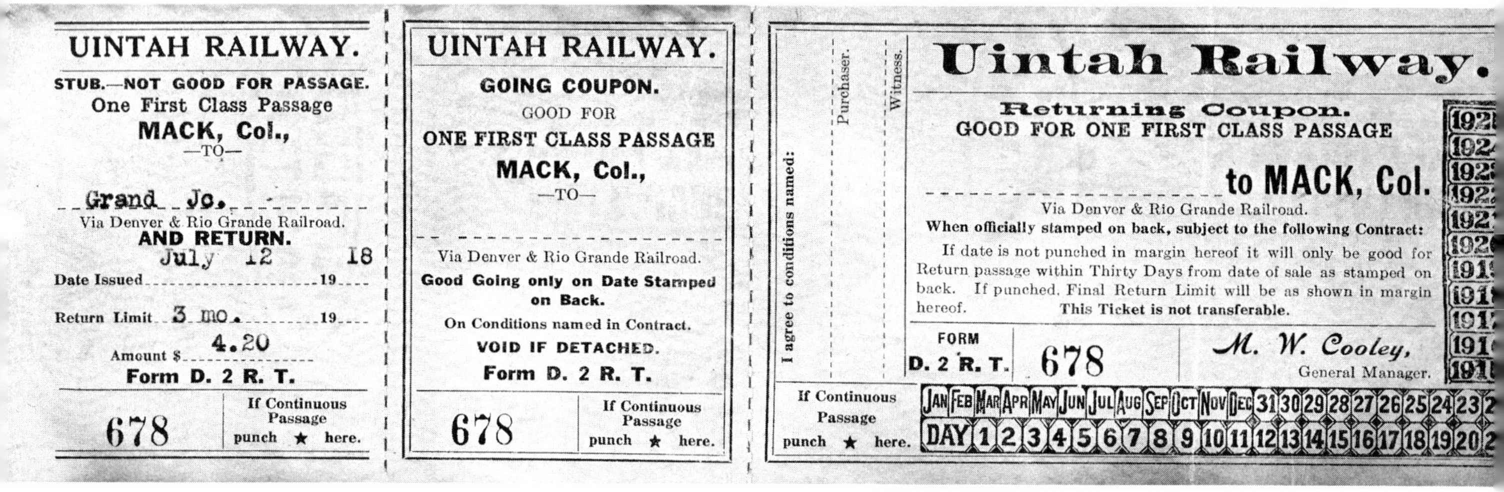

UINTAH RAILWAY.
STUB.—NOT GOOD FOR PASSAGE.
One First Class Passage
MACK, Col.,
—TO—
Grand Jc.
Via Denver & Rio Grande Railroad.
AND RETURN.
Date Issued July 12 19 18
Return Limit 3 mo. 19
Amount $ 4.20
Form D. 2 R. T.
678
If Continuous Passage punch ★ here.

UINTAH RAILWAY.
GOING COUPON.
GOOD FOR
ONE FIRST CLASS PASSAGE
MACK, Col.,
—TO—
Via Denver & Rio Grande Railroad.
Good Going only on Date Stamped on Back.
On Conditions named in Contract.
VOID IF DETACHED.
Form D. 2 R. T.
678
If Continuous Passage punch ★ here.

I agree to conditions named: Purchaser. Witness.
Uintah Railway.
Returning Coupon.
GOOD FOR ONE FIRST CLASS PASSAGE
to MACK, Col.
Via Denver & Rio Grande Railroad.
When officially stamped on back, subject to the following Contract:
If date is not punched in margin hereof it will only be good for Return passage within Thirty Days from date of sale as stamped on back. If punched, Final Return Limit will be as shown in margin hereof. This Ticket is not transferable.
FORM D. 2 R. T. 678
M. W. Cooley, General Manager.
If Continuous Passage punch ★ here.
JAN FEB MAR APR MAY JUN JUL AUG SEP OCT NOV DEC 31 30 29 28 27 26 25 24 23 2
DAY 1 2 3 4 5 6 7 8 9 10 11 12 13 14 15 16 17 18 19 20 2

RICHARD A. RONZIO COLLECTION

FOUR COMPANY HOUSES faced the Uintah Railway's mainline in this scene and gave the appearance of being part of a well-ordered company town. This picture was reproduced from a glass-plate negative that survived a fire at sometime during its existence, and it shows an early view of Atchee, probably taken ca. 1910. At the bottom of the view, in the foreground, was ground, was the town's saloon, a frame structure, with windows placed above eye level, with an entrance door facing away from the residences in the community for privacy. In the distant background, at the left, Atchee's original small school building was sitting in the mouth of the diminutive valley behind the settlement. This schoolhouse was complete with a boardwalk leading to the residences, and it had two outhouses, one for girls and one for boys.

FRANK A. KENNEDY PHOTO – COURTESY JOHN AND HAZEL CARSON

JOHN AND HAZEL CARSON COLLECTION

FRANK KENNEDY'S camera produced this view of Atchee probably about 1911. Several people were on the baseball field playing the national sport between the hotel and the enginehouse, as a long train of stock cars and gondolas sat on the siding. Housing already was becoming a problem, as witnessed by the white canvas tents behind the car shops and east of the wye. Notice the long timbers next to the tracks.

RICHARD A. RONZIO COLLECTION

ATCHEE, COLORADO—as seen from the hills south of town during late summer of 1912. Patrick Fitzpatrick, a conductor for the Uintah Railway, was standing in the foreground, with the residential portion of Atchee in full view. The larger white frame buildings were the original residences built by the railway for its employees and their families in 1904. As the railway grew and more residences were needed, additional ones were built at various locations around town. "Main Street" in Atchee was the yard of the Uintah Railway, which ran in front of the row of company houses that can be seen next to the carloads of sacked Gilsonite ore. A wagon road, shown in the lower left portion of the scene, headed out from Atchee to Carbonera and Mack,

but generally, most of the traffic that Atchee would ever see was on the steel rails of the narrow-gauge line. Spring water came from the mountains in the background, and running water was supplied to many of the houses at Atchee, but it was "hard water," containing many soluble minerals that had a tendency to corrode metals and make laundry stiff and cause white clothing to turn yellow. Rain water was a precious commodity for the laundry and cooking, and it was collected in barrels from the eves of many of the buildings. Under the window in the small dark-colored building in the middle of the picture, you can make out the water barrels and the makeshift rain-catcher along the side of the building, running down to the barrels.

FRANK A KENNEDY PHOTO – ARTHER HILLMAN COLLECTION

FRANK KENNEDY was quite energetic on this occasion, as he produced several different images from various points around town. This picture and several others were taken on the same day during late summer of 1912. The lady was used to add depth and perspective in the view, while gentlemen were used in at least two other views on the other side of town for the same effect. The boardwalk at this time was right up against the front of the houses with a large grass section going out to the tracks. The school building was on the back side of town, shown after it was expanded, complete with a bell tower and a new extension on the rear. A large expanse of sagebrush had been cut back to allow an area for the children to play. On the tracks several cars had been spotted, waiting to go toward Mack (to the right). The first car, from right to left, was a gondola, followed by a flatcar loaded with Gilsonite ore, another flatcar with unusual cargo and a stock car. The "unusual cargo" was a real money-maker for the Uintah in the fall. Carloads full of freshly cut wool were loaded at Watson and Dragon and shipped to Mack for transfer to the standard-gauge D&RG. The wool was shipped in huge burlap sacks that were stacked as high as possible. Wooden poles were put into the end stake pockets on the flatcars and the

large bags of wool were stacked in tiers with canvas spread across the top for protection against sparks from engines. You can plainly see one of the empty flatcars, with the large poles in the corners in front of the enginehouse. In the distant background, near the fading sun, was the large ranch house owned by the Kelleys. This was a large, two-story house situated in the broad valley above Atchee. An irrigation ditch, barely visible on the hillside above the sagebrush flat was fed by a small diversion upstream in West Salt Wash. Fields were kept in good order for a number of years; however, eventually raising cattle on the miles of open range turned out to be more lucrative than growing anything at Atchee. This large house burned down, and it was no longer visible by the 1920's. The Kelleys continued to ranch in the area from a smaller home up West Salt Wash. The Uintah's Bridge 28-A was above the lady's arm, crossing West Salt Wash. The saloon is near the middle of the view, with two windows high above the door level this side of the track. At the bottom of the wash, winding along next to the creek, was a well-used path, which led from near the section house to the saloon. This path was used by men to pass unseen by wives or management from the railway yard to the saloon under the cover of the deep wash.

FRANK A. KENNEDY PHOTO – RICHARD A. RONZIO COLLECTION

IN AUGUST OF 1912, Frank Kennedy climbed the hillside east of Atchee. Conductor Patrick Fitzpatrick was sitting on the rock at the left. In the distance, the Uintah's mainline track went upgrade from Carbonera, following West Salt Wash Creek. The building of the loading ramp to the top of the coal tipple necessitated the extension of the wye back farther toward the hill. A cut was made into the hillside, deep enough to allow an engine and a single car to clear the switch. This switch was a three-way stub switch allowing the car to be pushed down to the machine shop, onto the coal tipple, or back into the railway yard. The dirt from the excavation cut into the hillside was moved over around the small oil-storage shed at the end of the wye. The roof of a small building can be seen sticking up above the dirt. It is unknown what benefit came from the building being partially buried, but most likely it was for insulation, to keep oil and grease at a more even temperature during the hot summers and cold winters.

FRANK A. KENNEDY PHOTO – RICHARD A. RONZIO COLLECTION

A CLOSEUP VIEW of the railway yard at Atchee, taken from the previous photograph, shows many details about the Uintah. The little white building next to the hotel, at the upper left side of the picture, was the only store, and used by everyone in town. A train loaded with this year's harvest of bagged sheep wool was sitting on the mainline siding. Observe how crude wooden stakes have been installed in the stake pockets, allowing the wool to be stacked higher, and the canvas tarpaulins have been placed over the load to keep cinders and moisture off the load. The enginehouse and machine shop take up the space in the center of the wye. The original small boilerhouse was built on the east side of the machine shop and provided steam for the shop. The smokestack was built tall enough to help keep the smoke above the rooftops in town since the boiler was fired almost 24 hours a day. Notice the steam exhaust from the roof vent just past the smokestack. Sitting in front of the enginehouse was one of the Uintah's three-tank water cars. Notice the plank catwalk across the top of the tanks. Originally, these cars had two tanks mounted on each deck, but were rebuilt with three, as can be seen across the rail yards at the car shops where one of these cars is being rebuilt. To rebuild them, the original tanks had to be moved from directly over the trucks closer to the ends, allowing enough room for a new tank to rest inbetween. Look closely at the rebuilt tank on the front of the water car, in front of the enginehouse. The old tanks had tapered sides and had quite a few metal bands around them for support. The new tanks were not built to the same plans. They were the same diameter from top to bottom, but utilized fewer bands. Behind it sat one of the Uintah's small four-wheeled cabooses and engine No. 11, in front of the water tank. In the foreground were the company corrals. The stock pens were used primarily by local ranchers for sheep movements. The small building next to the loading chute was used for paint storage. Unfortunately, it had to be used at least once as a morgue, when a section hand froze to death on Baxter Pass during the late 1920's. The man was out in a bad snowstorm and failed to make it back to the section house at the summit of the pass. His frozen body was

brought down and kept in the paint shed until the next morning, when he was taken to Mack. The small square building beyond the two Shays contained sand that was poured into the locomotives' sand domes. This was used for ascending Baxter Pass, as well as for braking while descending. The two Shays were coaled the night before by the hostlers, using scoop shovels, and they were spotted on this section of track, ready to be dispatched the next morning. From here, the Shay geared engines could easily run down the wye track, back down to a train sitting on the mainline and head over the pass. Atchee was the distribution point for coal. Gondolas filled in Carbonera were brought to Atchee and placed on this siding. From here, they would be taken to the top of the coal tipple and dumped, or added to a train going over the "Hill" to Dragon, Watson or Rainbow. On the closest track was one of the Uintah's passenger cars and one of the "tool" or Bridge & Building Department cars. It was a cross between a boxcar and a caboose, having a cupola. Next to it was the pile driver and the remains of engine No. 10, from the wreck during the previous year. A man can be seen walking along the edge of the coal tipple, on top of the piles of coal. A gondola had been set out on top of the tipple track, and it had already been emptied. The gondolas were of the "drop-bottom" type, allowing coal to fall out the bottom of the car into the tipple below. As the tipple was filling near the top, a lot of hand work with shovels was needed to spread the coal out, keeping it from spilling over the edge. Things looked busy at the car shop, as a gondola, water car and boxcar had been set out for repairs. A track continued into the machine shop via the large wooden doors at this end. The building was long enough to house two cars at a time. This building was covered with a galvanized roof and walls, while the shop portion that was built on the south side of it had roll roofing and board-and-batten siding. Doors into the wood shop accessed the hard-parts inventory in the yard, and another door accessed the wood supply shed on the other side of the machine shop. The low-lying peaked roof just past the machine shop was where lumber and other supplies were stored out of the weather.

FRANK A. KENNEDY PHOTO – RICHARD A. RONZIO COLLECTION

JUDGE KENNEDY'S CAMERA was at it again on the same day in 1912—this time showing the railway yard at Atchee—looking northwest toward Baxter Pass. The grade on the pass dropped down to the lowest point on the horizon, at a steady 7.5-percent downgrade all the way across the mountain before disappearing behind the hill in the foreground. It reappeared from behind the same hill, headed left and dropped down to the large cut in the mountain at Morro Castle, where it reversed direction again, going behind the hill. The track finally appeared in the bottom of the valley and made its way down to Bridge 28-A and on into Atchee, 2,012 feet in 5.8 miles. The cultivated pastures in the bottomland of the valley were part of the Rogers Ranch. The small cabin, barn and white tents were at the site of this small homestead. In March of 1915, Mrs. Josephine Rogers became the postmistress for Atchee and retained this position until the early 1920's. After being in the old store building and the hotel, the post office was moved to the Rogers Ranch in 1915. This was an unpopular move with the residents of Atchee, who now were forced to walk up the track to get their mail. The mail was delivered via the Uintah at the platform near the water tank, and Mr. and Mrs. Rogers would take it to their home for dispersal. Mrs. Rogers took her job a bit too seriously, and on several occasions, she got into heated debates with other ladies from town, who had to help carry their packages to Mrs. Rogers' house just to sign for them.

CHRIS H. UEBERROTH COLLECTION

IN THIS ENLARGEMENT, two snowplows were visible, one sitting in front and one behind a gondola, on the spur near Bridge 28-A. In the foreground, at the right, was one of the Uintah's tool cars, complete with a cupola and platform at this end. Sitting to the left of this car was the Uintah's newly built pile driver. Stacks of railway ties and bridge timbers were sitting on the ground next to the pile driver, on top of which sat the large wooden poles for the pile driver's hammer slide. Consolidation No. 10 was sitting at the end of the next spur, at the left. After being pulled from the mud the previous year, at the accident at Bridge 1-B on October 1, 1911, there was not much left of the exterior. The cab had been crushed and sheared off in the wreck, and the smokebox door had been opened to allow the inside to dry out and to provide access to the interior of the boiler for inspecting the damage and preparing it to be rebuilt. Engine 10's tender was behind the engine, pushed up behind the boiler backwards. A pile of bridge timbers had been stacked on planks, waiting to be utilized during the next series of flash floods in the canyon. The coal tipple is visible, with a gondola spotted on the top of it. The track to the top of the coal bunker branched from the three-way switch at the end of the wye and traveled up a dirt fill to a small trestle, which took it up to the top of the tipple. Coal had to be shoveled into the tenders of the engines by hand. The top of the coal chute was on the level, and the tracks on both sides went downhill toward the mainline. In this way, no matter what the level of the tender was, it could be moved up or down the incline to facilitate the right height for coaling.

INTERSTATE COMMERCE COMMISSION INVENTORY – 1919

ATCHEE ROUNDHOUSE, BOILERHOUSE, MACHINE SHOPS, CAR SHOPS

(Unless otherwise noted, all machinery was belt-driven)

–Commercial Electric Company No. 3305, Electric Generator, 125 volts, 10 amperes, 1,200 rpm.
–1907 Nagle Engine & Boiler Works locomotive-type steam boiler. 57" diameter 22' long
–Westinghouse 9½" Air pump
–Atlas steam engine, 8" x 12"
–Fairbanks Morse Company Gas Engine, 8 horsepower
–1919 Flanders Machine Company, portable boring bar 8" to 2" capacity
–1907 National Machinery Company, 1½" single-head bolt cutter
–1907 Cincinnati Machine Tool Company, 32" Drill Press
–1907 Lodge & Shipley, Engine Lathe 16" x 10'

JOHN CARSON COLLECTION

THIS VIEW OF ATCHEE was photographed as you looked southeast, showing most of the town just a few years after construction. The small schoolhouse was in the mouth of the large canyon at the back of town, with two windows on the northeast side, a small front door and the outhouse in back. The white company houses and the dark-colored hotel were the only real residences in town at this time. The white house this side of the track was constructed as space for more employees during the same year as the schoolhouse. Next to it was a large white canvas tent. Many early residents lived in tents like these—some for as long as a year—waiting for more permanent structures to be built. On the mainline sat two boxcars, a gondola and one of the Uintah's original small four-wheel cabooses. Just this side of the cars, in the small wash below the level of the track, was a fancy "two-hole" outhouse. Built on pilings above ground level, it was for passengers who were waiting for trains to leave this location. There was no official depot at Atchee and no "services" for passengers. A wooden staircase was designed to take the traveler from track level down to the privy, which was complete with a small louver on the roof. In the foreground was the Uintah's section house. This was one of the first buildings constructed, and it was used by the first railroaders while construction continued along the line, even before the residences were completed. Two small "dugout" residences can be seen behind the section house in this view. They were complete with low ceilings, dirt roofs and stove pipes sticking up in the corner.

INTERSTATE COMMERCE COMMISSION INVENTORY – 1919 – Continued...

–1907 Lodge & Shipley, Engine Lathe 42" x 14'
–1917 Searchlight Company, Oxy-Acetylene welding outfit
–1907 William Sellers & Company, Planer 30" x 30" x 8'
–1907 J. F. Schaffer Manufacturing Company 150-ton driving-wheel press
–1907 Steptoe 20" Shaper
–1914 Greenlee Brothers, Post-borer, 3½" bit, 20" Stroke
–1914 Hall & Brown 8" x 24" capacity wood planer
–1914 Sidney 32" Band Saw
–1914 Cresent Machine Company 16" Universal Saw

Also listed were: 1 blower and piping, 1 forge with bin and vat, 1 steam hammer, 1 face plate, 2 runaways and trolleys, 1 flue-welder, 2 anvils, 8 vises, 1 air reservoir, 2 grinders, and miscellaneous small tools.

JOHN AND HAZEL CARSON COLLECTION

A CLOSER VIEW of a portion of the last photograph, taken about 1911, shows the engine to be the Uintah's second No. 11. Fresh from Baldwin Locomotive Works, she replaced the old secondhand Denver & Rio Grande engine named the *Tomichi*. At the time this photograph was taken, she was the newest locomotive on the roster. **UINTAH RAILWAY** was precisely and boldly lettered on the tender. The little 2-8-0 was ready to back out onto the mainline.

THE SECOND No. 11 was purchased during February of 1911, in conjunction with the extension of the railway north from Dragon to the mines near the new mining camp at Rainbow and the northern terminal of the railway at Watson. Traffic over the line had been growing steadily for the past few years, and the volume of traffic expected during the construction and after the new mines were opened, warranted "faster" power on this side of Baxter Pass. The second No. 11 was ordered using the same plans as No. 10. Of course, a few details were different, but they were basically the same. Engine 11 had a large base on the steam dome, unlike the one on No. 10, which had the same diameter from top to bottom. Also, No. 11 had two panels under the cab windows, whereas No. 10 had only one. This picture was taken at Atchee ca. 1918, where the 2-8-0 was sitting on the mainline siding, with the locomotive heading downgrade in the direction of Mack. The hatch on top of the engine had been opened to help ventilate the hot cab. Notice that the original smokestack had been changed to a capped stack.

JIM McCOY COLLECTION

FRANK MANCUSO COLLECTION

THE SECOND ENGINE 11 was sitting in front of the water tank, ready to head out on her run back downgrade to Mack. Like all the 2-8-0's, the grade on Baxter Pass was too steep to surpass, so No. 11 stayed on this side of the pass, not going any farther north than Bridge 28-A.

FRANK A. KENNEDY PHOTO – ED AND NANCY BATHKE COLLECTION

SPECIAL TRAINS for excursions to the top of the pass always were popular and fairly common, especially during the first seven years of the Uintah's history. Here, everyone was in their finest clothing, with children waving American flags in celebration of the Fourth of July—as the train was getting ready to head out toward the cool mountain air at Baxter Pass and Columbine. Coupled behind one of the old secondhand combines, No. 52, was flatcar No. 111, turned into a makeshift observation car. Sections of large, recently hewn trees had been used to build the uprights, which had been wedged into the stake pockets for support. A canvas tarpaulin has been strung across the top for sun-and-rain protection, and wooden planks had been installed for seating. Nearly everyone in town would go on these outings. On the Fourth of July, the combined populace of Mack, Carbonera, Atchee and Dragon could tax the passenger capabilities of the railway. This photograph was taken near the enginehouse, looking north. The water tank was just out of view, at the right. Notice that extra drive wheels had been placed near the mainline.

(Continued from Page 173)
chee, along with the Uintah Railway and Garfield County, combined their efforts to build a new schoolhouse on the west side of town. This building was a much more modern facility, with running water, a furnace, a large basement and a separate small apartment for housing the teachers, known as the "teacherage." The old building was used by the school and the railway for storage, mostly old paperwork. Within a few years, the old unoccupied and rarely used building burned down. Rumors were whispered about the coincidence of some missing records concerning school funds and the fire, but nothing more than small-town rumors ever came of it.

The new larger building was built on the south end of town, at the base of the large mountain to the southeast. A concrete foundation and basement were poured, along with a concrete stairwell up to the level of the main floor. Two large school rooms and a large auditorium-style room toward the back, made up the main level. Another large room was in the basement, and there also was a large coal-burning furnace in the basement, underneath the stairwell. It was here that another fire started around 1930 or 1931. Like any structure that caught on fire in Atchee, the school building burned to the ground, as the helpless citizens stood by in despair. Residents were thankful that no one was in the basement at the time, because the location of the start of the fire would have surely trapped them inside.

For over a year, school was then held in one of the employee's houses. The community rallied around the theme of rebuilding the school. It was during the Great Depression, and money for the small isolated community was hard to find. Finally, with money from Garfield County, the Uintah Railway and funds from the citizens themselves, there was enough money put together to rebuild the school. Employees volunteered not only their money, but their labor. All of the labor was provided by the men from Atchee working around their rail-

ARTHER HILLMAN COLLECTION

THE ENTIRE SCHOOL often would ride aboard the train to Baxter Pass on a field trip. Here, the children were standing in front of combination car, No. 50, getting ready to travel in the passenger car to the top of the pass, or over to Lake McAndrews. Sometimes, the students would go as far as Dragon or Watson; however, on most occasions, they would get off on the pass and wait for the returning train to Atchee. The children were standing on the wooden platform at Atchee.

way schedules. The basic plans from the old building were used and improved on, and the new building was built within a few months. School at Atchee was for most years a two-room classroom with two teachers. By the late 1920's, class size had decreased to the point that one teacher could take care of the entire student body.

Teachers were brought in from around the country because the local school board was willing to pay more for the position than most teachers made in large cities. The school board was a group of men from Atchee who oversaw the business end of School District No. 39. Advertisements were run in certain circles, and qualified individuals were selected for interviews. Using their railway passes, a couple of the school-board members headed out across the country in search of a new teacher. As far away as the East, young ladies were selected for the positions at Atchee, and for a few years, Carbonera and Wendella. A "teacherage" was located next to the school and was basically a very small shack-style house. At one time, a member of the school board was able to get a relative a job as a housekeeper for the teacher, and this older lady helped with not only the housework, but in doing janitor work in the school. One big requirement for the job was the ability of the teacher to play the piano. Being the only large public building in town, the school was a natural location for dances and socials. The official reason was given that the children should be instructed in the basics of music, but the real reason was that the piano player in Atchee was needed for social functions, almost as much as bootleg "booze"!

Owing to its large size and public space, the school at Atchee was the main source of social functions. The daily gossip at the general store, the mail at the post office, or the comings and goings of travelers at the hotel, were all interesting and needed, but the school was where the special events took place. School plays were regularly put on around the traditional themes of Thanksgiving and Christmas. During the 1930's, Frank "Judge" Kennedy would drive a motor car or come in on the train and set up his small motion-picture theater in the school. This eventually was done on a weekly basis, and was anticipated by everyone, especially the children. Dances and socials also made great use of the space at the school, and many times, the piano would play late into the night, as the railway employees, ranchers and sheep men from around the area would converge on the small

FLORENCE REYNOLDS COLLECTION

SCHOOL AT ATCHEE was always interesting, and for most students looking back on their time there, they received a better education than most other schools they could have attended. Children received individual attention, and they experienced things earlier in their education than children in other towns. Movies and talks from Frank Kennedy, typing classes from Mr. Beard, visits from traveling preachers and the constant overseeing by teachers in a town where skipping school was impossible, added up to good, solid education that took many of the children far when they went to high school in Grand Junction or Fruita. Mrs. Gooch, on the left, and Mrs. Burgess, on the right, were proud of the Atchee school, as they stood for their picture before sending their pupils off to the championship spelling bee.

THIS YOUNGSTER had been posed for this photograph while sitting in a child's "covered wagon." The picture was taken next to the corner of the new school building.

RODGER POLLEY COLLECTION

JIM McCOY PHOTO – RODGER POLLEY COLLECTION

TWO TEACHERS STOOD next to the boy and girl winners of the spelling bee in Atchee. These two children would get to ride on the Uintah Railway to Mack, and then, via the D&RGW, they would travel to Glenwood Springs, Colorado. In the background, No. 12 was waiting for water cars to be filled prior to this trip to Mack. The plank passenger platform had been removed.

ROBERT C. BOULTER COLLECTION

THE BELL TOWER of the old school at Atchee was located just above the front entry way. A long rope hung down from the bell and was tied off to the wall in the hallway. One of the student's favorite tricks was grab the rope and ring the bell on their way in or out of the school—without being caught by the teacher. In the foreground were several small outbuildings used for storage or as privies. This picture was taken during the early 1920's, after the new school had been built, and this building was being used for storage of records and paperwork from the railway and the school.

ROBERT C. BOULTER COLLECTION

ATCHEE'S NEW SCHOOLHOUSE was built about 1920, a part of Garfield County School District No. 39. This building burned and was rebuilt ca. 1930, using the same exterior plans, but with some changes on the inside. This view was recorded from the residential part of town, looking southwest, down West Salt Wash Canyon, toward Carbonera and Mack. The smaller building in the foreground was known as the "teacherage," and whoever was employed as the teacher resided in this small house. The boardwalk extended from the front of the building to meet the boardwalk in front of the company residences. Many times, when it was rainy and muddy outside, the children who lived on the other side of town would "take the long way home" by means of the boardwalk rather than get muddy and dirty by walking through the middle of town on the dirt (and sometimes muddy) road. Notice the large children's swing placed next to the building.

building. The school was also used as religious center at times. During the early years, traveling ministers often would come by train to hold services on Sundays, in Atchee and the other towns along the Uintah, for several weeks at a time. This practice became more uncommon by the 1930's, but at one time, Atchee had an Episcopal women's group that met on a weekly basis.

In 1922, a young Eastern girl, Elizabeth Angus, was hired as a school teacher. Somewhat well-to-do, as well as college educated, she was appalled by the simple living conditions and isolation at Atchee, and she started her own Sunday School for the children in the school building. Her enthusiasm ran out after a couple of summers, but by that time, she was married to a Uintah employee and would stay there for the next 13 years.

The newest school building outlived the Uintah. By the time the scrappers were tearing Atchee apart, Garfield County was not prepared to let the school be torn apart for its lumber. With no town, no children and no teachers, the building sat vacant for nearly a year before the powers that be finally let it be torn down.

In such a small town, recreational activities were somewhat limited, but the things that were done were done well. Social get-togethers were designed around any kind of holiday or special occasion, allowing the townspeople to entertain themselves.

Dances at the schoolhouse were the most remembered and participated-in activities that Atchee had. Not only was the schoolteacher required to be able to play the piano, but there was a regular band comprised of employees and their wives. The band was a long-standing tradition in Atchee. Even as employees came and went, there was always someone else who wanted to join in on the fun. The schoolhouse was used by the band to hold the dances, and this was a big part in the redesign of the interior of the school after it burned down in the 1930's. It was rebuilt, with removable walls in two classrooms, to allow the auditorium to hold more people. As the townsfolk would make their way to the school, many of them would bring their children. The school chairs were moved out of the center of the room, with the backs of the chairs facing away from the walls. As nightfall came and the children's bedtime approached, the parents would simply make up small beds on top of the chairs and put their children to sleep in the "dance hall." Many of these dances would last until late at night, as many folk came from all over to participate in the dances.

One young employee of the railway remembered learning how to dance while still just a teenager, and how that helped him out when he moved to Atchee, after getting a job with the Uintah Railway. "A young lady was in town visiting her parents, who worked for the Uintah, when we were having

RODGER POLLEY COLLECTION

AT LEAST ONCE each year, the pupils were posed at the school for their picture. The front stairs provided the place for this view, taken in the mid-1920's. Behind the teacher was access to the coal furnace in the basement. It was this furnace that caused the building to burn down during the 1930–1931 school year. To the right of the staircase was a large coal bunker, which had to be loaded by hand during the winter. Notice that a door for a coal chute was behind the teacher.

town dance. I knew how to dance. Well... let me just say she didn't go to the dance with me, but she left with me. We got married not long after that." The crowds were pretty big for the dances, and by the 1920's, as more and more people bought automobiles, there usually would be a large group of miners and their families who came up from Carbonera, as well as ranchers and sheepherders who were in the area.

The band was a mixture of musicians with different talents, who met together for fun. Many of the other townspeople were not involved in the band, but knew how to play instruments, also. At least one family had a player piano, which was used over and over again as a social centerpiece in this quite town. Radios never got a big hold in Atchee, as there was poor reception in the isolated valley, but the radios that worked during the 1930's usually were tuned in by more than just one family.

In Atchee, as well as the rest of the nation, baseball was a favorite summer pastime and families in Atchee were avid participants. The earliest baseball games were played in the open space between the residences and the railway wye. This field was overtaken by the car shops and employee residences during the teens, and the ball field was moved to the east side of the tail track of the wye. This field was nothing more than a rough diamond, kept clear of sagebrush and weeds. Baseball was a big part of summer life for many railroaders, and much pride was associated with the teams. Small towns all over the country sent out their finest men for competition after competition, which often seemed more important than the jobs they held and town they played for. It was well known that if a man could pitch or hit, it was a big plus for him when looking for a job.

A number of employees who were hired by the

ROY ENO JR. COLLECTION

THE NEW SCHOOL, as originally built, had a large basement, with a separate entrance on the southwest end of the building. This picture was taken in 1927, and it portrays children from Atchee in front of the entrance to the basement. When the building was rebuilt after the fire, the foundation was reused, but the basement was filled with dirt, providing only a small crawl space.

RODGER POLLEY COLLECTION

TAKEN FROM NEARLY THE SAME PLACE as the previous picture (on Pages 180–181), this view shows Atchee during the late 1920's. Many more residences had been built, some had been moved or torn down, and a many had been added-onto. The back side of the schoolhouse is at the bottom of the picture, and just above it and to the right were two garages, used to store the automobiles that saw infrequent use

between Atchee and the Grand Valley. The winding horse trail through town had also changed, and it was now a well-used thoroughfare for motor cars. This was as large as Atchee would ever become, although a few more houses might have been built at a later date. However, the work force for the Uintah Railway would never become any bigger than it was during the mid-1920's.

ROBERT C. BOULTER COLLECTION

THIS SCENE was photographed from near the top of the ridge where the townsfolk of Atchee maintained a flag pole. It was accessible by way of a lengthy hike in the area of Bridge 28-A. The enginehouse, machine shop and water tank are visible in their tight arrangement at the wye. Long strings of flatcars and a caboose lined the mainline. Luton's store, located this side of the track, stood beside other residences, and the icehouse can be seen next to the lower string of flatcars.

Uintah Railway during the teens and 1920's were thought to have a number of talents not necessarily associated with their railroading skills. Teams were put together by many small towns, and most of the Uintah towns, including Atchee, had their own. Baseball games were hosted in Atchee on many occasions, as teams from Grand Junction, Fruita, Mack and Dragon would make their way via the train to take part. Most times, the games were not associated with any league or ranking system, but were held for the fun of the participants and spectators. By the late 1920's, many people in Atchee owned automobiles and spent most of Sundays in Grand Junction, Fruita or Mack, either watching or playing baseball.

Sunday was also a day for observing religion by many people, and a number of townsfolk would leave in motor cars on the weekends to go to church in the Grand Valley. There never was an official church organized in Atchee. The school building was used as a church on an occasional basis, as traveling preachers from different denominations

FLORENCE REYNOLDS COLLECTION

SIX-YEAR-OLD MARY BURGESS was climbing the long trail to the flag pole in this photograph taken during the 1920's. Fourth of July celebrations at the flag pole were cause for a late-afternoon start up the hill, as the daytime temperatures during the middle of summer were too hot for strenuous effort, the cooling shadows of the late evening made for an easier climb. The climb was quite a hike, starting near bridge 28-A, and climbing nearly 800 feet to the top of the ridge west of Atchee. This ridge line started here and ran west up towards the top of the pass. This picture was taken looking northeast, down in the valley, where the Rogers Ranch was located. Atchee was just around the corner, at the right.

RODGER POLLEY COLLECTION

AS SEEN FROM THE HILL west of town a few years later, Atchee had well laid-out yards and fences. The original construction of company houses contrasted greatly from the random placement of the rest of the newer houses in town. The original houses had been laid out by the railway surveyors and professional contractors, complete with backyard fences and straight boardwalks. Other buildings in town, along with the schoolhouse, were placed in relation to the surroundings. The new school building was to the far right in this view. A large playground and field had been cleared. Notice the boardwalk crossing the road and connecting with the old boardwalk in front of the houses. The road to Atchee from Carbonera came up the valley. It split in two to the right of this view, with one side continuing toward Atchee, while the other side crossed the railway tracks and made its way up to Luton's store. The main road into town followed the path of the old wagon road and curved around behind the main lots to a location near the end of the wye. From this point, the road was not maintained, as it crossed the tail of the wye and continued on through the sagebrush field east of town. From there, it went on up to a ranch in the canyon above and to Rogers' Ranch on the other side of West Salt Wash Creek. At this point, the road all but disappeared, except for a rough track that continued on up to Baxter Pass. Although this route was not totally inaccessible, few people tried to drive their cars up this road. For the most part, it was only accessible by wagon and horses, in a time when they generally had been replaced.

would frequent Atchee during the summer months. Up into the 1920's, many preachers would arrive by train in the summer to preach in the small community. A Methodist preacher from Grand Junction would often spend a whole week in Atchee and then go to Dragon before returning home. One Episcopal preacher from Fruita made it a point to come to Atchee at least twice a year, and during his visits, he would spend an entire day at the schoolhouse teaching the children Bible stories. The preacher had a Southern accent, with a deep voice, and the children were spellbound by the larger-than-life stories of Joseph, Noah and the prophets. In the days when school prayer was virtually required, this practice was not looked on with disdain by the parents but instead encouraged.

Holidays created other social events that kept the town busy with preparations for weeks in advance, and full of gossip for weeks afterwards. The Fourth of July was always a big celebration in Atchee. The most remembered event was the special trains that were operated for the celebration. Passenger cars or flatcars with poles and canvas tops were incorporated to haul employees and their families to the top of Baxter Pass for picnics and socials. Employees of the railway would volunteer

ROY ENO JR. COLLECTION

THE TOWN DUMP at Atchee made for an unsightly mess on the east side of the wash as of 1926. A natural place for trash to be deposited, out of sight and out of mind, the deep wash made by West Salt Wash was used to dump not only the town's trash, but also was a convenient place to empty the railway's ash pits. This picture shows the worst part of the dump, behind Luton's store (the black building on the left). Also located here in 1926, was another one of the Uintah passenger cars being used as a residence. The long white building was one of the old Pullman sleeping cars, with a front porch and a privy on the back end. On the far right was Atchee's newer school building. This was built to house the larger population of school children during the early 1920's.

JIM McCOY COLLECTION

A YOUNG GIRL and her puppy were standing in the front yard of one of the residences at Atchee. In the background were the railway yard and icehouse. The original construction through the yard basically was one long cut. The extra dirt from the cut was thrown out on the west side of the tracks, forming a long dirt bank next to the right-of-way, visible in front of most of the company houses. This was a natural place to build the icehouse, using the freshly dug soil to build the foundation. The walls of the icehouse were built with railway ties and thick bridge-style timbers for insulation and to hold the dirt back. To the left and right of the front door, the dirt was held back by cribbing. The soil was piled up nearly to the level of the roof, allowing the inside of the building to stay cool even in the summer. The building had a metal roof, and a small vent on the top allowed the summer heat to escape. Ice was brought in from Lake McAndrews in boxcars, and it was stacked inside the cellar-like building in tiers, with sawdust in between. The residents could utilize the ice whenever they needed it and would use a pair of ice tongs to grab some ice and take it back to their homes.

RODGER POLLEY COLLECTION

UTE INDIANS WERE a common sight in Atchee. Not far away from the Uintah Indian Reservation, Atchee was a natural place for the Indians to visit on a regular basis. Luton's store was usually their most common place to stop, and many stories were generated over the years by the clash of two totally different cultures. Everyone in town loved the Indians, and no one ever remembered a harsh word or tense moment from the always polite, peaceful and happy Utes. On one occasion, a small group of Utes appeared on the other side of the wash from Luton's store and proceeded to set up camp. Ruby was known as the person to see if something medical was needed, and late that night a young Ute "buck" came into the store and into the personal quarters of the Lutons. Not able to communicate in English, it was finally figured out that he wanted her to go with him as his squaw was having problems in childbirth. Ruby went with the Ute to their small tent and helped deliver the baby. On the following morning, when Ruby got up and went to see how they were doing, all she could see was the family on horseback crossing the horizon, just going out of sight. Mr. Luton also had his encounters with the Indians. The Ute squaws were definitely second-class citizens, and on one occasion during the summer, a group of Ute men bought a large quantity of watermelon and proceeded to eat it without giving the women any. This upset Metters so badly that he had several watermelons cut up and given to the squaws for free. In turn, they were grateful, and from then on, the Ute women always greeted Mr. Luton with large smiles and waves, something that no one else in town received.

RICHARD A. RONZIO COLLECTION

AUGASSI, AS HE WAS KNOWN, was the one of the most well known Utes in the area. He was somewhat of an outcast in the Ute tribe and had been branded a coward and forced to wear women's garments when he was a young man. It is not known what exactly it was that he had done, but reports varied over the years from just refusing to go to fight, to actually running away during battle. He hated to have his picture taken, and most pictures of him are just like this one, turning way from the camera near the cattle guard on the other side of Bridge 28-A. Augassi spent a lot of time in Dragon, Utah, and in the Uintah reservation, but from time to time, would wind up in Atchee with whatever band he was with at the time. Being a curiosity to the adults in town, he was looked on with fear by most of the younger kids in town. Taunting and constant harassment by not only his tribesmen, but the white man had made for long hard life, and it could be said that he was a little bit on the grouchy side when it came to the pranks of youngsters in town.

FREDA AND FRANCIS RIZZARI COLLECTION

IN THIS SCENE, most of Atchee's school children had posed with Metters and Ruby Luton (at the far right), along with some the children's parents, in front of the Luton's store. A post office sign had been hung over the separate entrance to the post office, while the door to the right went into the general store. The Solitaire Coffee logos, both on the sign and the windows, were part of a promotional campaign by the coffee-maker to co-advertise. The signs were paid, in part by Solitaire, to advertise their coffee, and at the same time, created a well-made sign for the business. This was a very successful campaign, in this area at least, because—with the exception of the name of the establishment—the general stores in Dragon and Watson, Utah, and Rangely, Colorado, all sported the same signs, with the two large Solitaire logos on the front.

TOTAL CHARGE IS FIGURED TO NEAREST FIVE CENTS

NAME M. A. LUTON — $ 3 c 00

MAIL ADDRESS ATCHEE, COLO.

6 Exp.	8 Exp.	Pack	Cut Film	Neg.
6 Exp.	10 Exp.	Pack	9 Neg 130	Neg.
6 Exp.	12 Exp.	Plates	Neg.	Neg.

	SPECIAL	Pack No.	Print No.	Order No.
Developing	Post Cards			Ex. Col. 276
Printing 11 each			Cutter No.	
Received JUN 24 1929				
Finished JUN 24 1929				

KODAK FINISHING

Order Checked B — Double Checked

M. A. LUTON
ATCHEE, COLO.

RODGER POLLEY COLLECTION

M. A. LUTON
General Merchandise
Dry Goods, Notions, Groceries and Meats

Atchee, Colo., 8-15 192

M

Reg. No. Clerk ACCOUNT FORWARDED

1		
2		
3	2 Bread	3 00
4		
5		
6		
7		
8		
9		
10		
11		
12		
13	23	
14		
15		

Your account stated to date. If error is found return at once.

RODGER POLLEY COLLECTION

THE TEXACO FUEL PUMP was in front of the Luton's store during the late 1930's. An employee's boarding house was in the background, and the front end of one of the Pullman sleeping cars, which had been converted into a residence, complete with a screened-in porch, in place of the end platform. This old-style gasoline pump would not be seeing much service for awhile, as snow all but ended automobile traffic in and out of Atchee until spring. The reason they were called "pumps" is that was what you literally had to do to "gas up" a car. If you wanted three gallons of gasoline, you had to pump the fuel from a container underground up into the glass cylinder, then place the hose into the car and let gravity feed the gasoline into the vehicle.

their time, and the Uintah would volunteer its equipment, to make the special Fourth of July runs. This practice was abandoned in the 1920's, as the management trimmed fat from overhead and operations, but July Fourth continued to be a big celebration. On top of the tall, pointed mountain west of Atchee, a flagpole was erected during the 1920's for the Fourth of July celebration. Townsfolk, laden with heavy picnic baskets, would hike to the top of the mountain, looking down on the town and out over the Bookcliff Range, toward the Grand Valley. This secluded eagles' nest became the favorite hangout for teenagers in the 1930's, and it was one of the few safe places to smoke cigarettes without being caught by adults!

On one Fourth of July, in the early 1930's, a group of machine-shop employees decided to celebrate with some black powder, commandeered from the shop. A small barrel of the stuff was sifted out across the railway yard and up along the boardwalk to the hillside east of Atchee, behind the wye. A match was put to it and the show really began. The fire leapt along the open ground and across the railway yard, as the townsfolk watched. However, instead of quickly burning out at the end of the trail, the powder ignited a batch of dried grass and weeds. Before long, a wildfire was in full swing, rolling across the bottom of the hill and threatening homes and railway equipment near the end of the wye. The men worked feverishly to put out the flames, and finally did after a few hours of hot, exhausting work. As can be expected, there were questions that had to be answered about that incident, and several good shop employees almost lost their jobs.

Atchee was a real community. Although the amenities of city life often were vacant in this small railroading town, everyone who ever lived there had nothing but fond memories of the time spent in this small isolated town in western Colorado. The common employer (the Uintah Railway), the many dances and socials, and the everyday strides of everyone to make the best of everything, created a very loyal and helpful community that was there for everyone's problems and needs.

Of special interest was one occasion when a young boy was injured and the community and employees of the railway worked together to help him out. Jim McCoy, nicknamed Shanny Irish, was sledding on the steep hill behind the school building

CLIFF HAYCOCK COLLECTION

ATCHEE'S UINTAH HOTEL was the smallest of the three hostelries built by the Uintah Railway in 1904. More a boarding house than a hotel, the structure was used mostly by single employees rather than the traveling public. While Mack and Dragon were natural stay-over points for travelers in and out of the Uintah Basin, Atchee was not a destination for most travelers. Rather than staying the night, a change of motive power and a quick stretch of the legs was all that most people saw of Atchee. The hotel was run like the other hotels in the early years, complete with prepared meals and company china at every table. As the years went by, the hotel saw infrequent use by management from the East, snowbound travelers waiting for Baxter Pass to be opened, and train crews stuck in Atchee for the night instead of their homes in Mack or Dragon. The company painted the cottages white from the beginning, but the hotel originally was a dark color, although by the mid-1920's, it had also been painted white. The railway hired a painter, and every two years the cottages would get a new coat of white paint, as well as for the hotel during the later years. This photograph shows the hotel during the early 1930's, when it was known as the "Luton Hotel," named after Ruby Luton, proprietor at the time. Mrs. Luton was a charismatic individual, and whether or not there was an official name change, the hotel was known by all as the Luton Hotel. Two tall masts for a radio antenna were in view, at the left.

with a large group of other school children. The long sled would be pulled up the side of the hill. At the top of the hill everyone would pile on the sled for the ride down the slope. At the bottom of the hill was a fence with a gate in it that the sledders had to maneuver through. On one occasion the sled hit one of the large fence posts, breaking the leg of young Shanny. Company doctor Montgomery happened to be staying at the Atchee hotel that evening, on his weekly rounds of the railway, and was quickly summoned to Jim McCoy Sr.'s house to examine the boy. In great pain all through the night, and with a badly broken leg, the doctor decided that Shanny would need medical attention in Grand Junction or Fruita. The problem was that snow on the roads had made the trip to the Grand Valley impossible for a car. With no scheduled train on a Sunday, calls were made to Mack to get permission to run a special to Mack. The OK was given and Shanny was placed on a stretcher and bundled up for the ride to Mack. The train crew donated their time for the special, and a single engine and passenger car were taken down to Mack with the boy. He would later recover fully, but sledding on the hill behind the school was strictly off limits from then on.

When Metters and Ruby Luton moved to Atchee around 1920, the only store that Atchee had ever known was located behind the hotel in a small one-room company-owned store. Selection was minimal, and the prices were high. It has been lost to history as to exactly why the Lutons decided to move to Atchee, but once they arrived there, they would be there for the remaining years of Atchee's exist

RODGER POLLEY POSTCARD COLLECTION

UINTAH RESERVATION INDIANS, like Chief McCook, were a common sight in Atchee during the summer months, as this Ute tribe utilized the grazing land and cooler climate of the mountains west of Atchee. It seems as though Chief McCook took a liking to Ruby Luton, intrigued by her red hair. While visiting the Luton's store, Chief McCook entered the store all dressed up in his best clothing, and he seemed to be only interested in talking to Metters Luton. After a few minutes, it was clear that he had come in to trade for Ruby! Much to Ruby's chagrin, Metters decided to play along and find out how much she was worth to the chief. When bidding got to two hogs, Ruby decided to leave the store until Chief McCook had departed for other parts! Later, Metters told Ruby the bidding got all the way up to *four cows and two horses!*

RODGER POLLEY COLLECTION

THIS SMALL MULE DEER was one of the more unusual pets in town, raised from a fawn in Atchee. The deer was hand-fed by Roy McCoy and grew up quickly. He was the talk of the town, as he was friendly and tame, and all of the children and most of the adults in town were amused by the young deer. It stayed around for nearly two years before it started to cause more trouble than it was worth. First, there were ever-increasing disappearing flowers and shrubs. Then, the young deer made its way into the back of Luton's storeroom through an unlocked door, where it proceeded to eat all the peanuts—much to the Lutons' dismay. The final straw came when the young buck started to grow horns and proceeded to butt anyone who stood still. The outraged mothers in the community brought immediate attention to the problem. Everyone had grown rather attached to the deer, and no one wanted to put it down, so it was decided to try and relocate it. It was tied up and placed in a northbound freight train. At Lake McAndrews, the deer was untied and set free in the wild. That approach was tried twice, with the same effect: The deer showed up back in Atchee within several days. This picture shows the ultimate solution. As General Manager Walter Rader and Scott Schafer looked on, the rope reins to the deer were handed over to a couple of Ute Indians. As everyone in town knew, the Indians knew how to take care of a deer, and he was never seen or heard from again.

FLORENCE REYNOLDS COLLECTION

A FAMILY GET-TOGETHER on the front lawn of Robert Hester's house at Atchee, Colorado, was held in July of 1938. From left to right, the family members are: Ed, Stewart and Bernice Boulter, Florence and Robert Hester, Johnnie and Anna High, and Bob Boulter. By this time, the trees and shrubs in front of the company houses had reached maturity and really changed the look of this small mountain desert town. Instead of dry patches of desert plants and sagebrush, they now were well-groomed, with shaded lawns in front of most of the houses in town.

IT WAS MAY 28, 1931, at 6:40 a.m., in Atchee, Colorado, and you are looking south in front of the hotel and cottages. The hotel's boardwalk ran from the main boardwalk, at the right, up to the hotel's back wing. The caretakers lived in the back wing of the hotel (at this time), while guests used the front steps shown in this view.

DAN PRICE COLLECTION

ROBERT C. BOULTER COLLECTION

THE FOUR HOUSES built along the main track were all constructed to the same plans. Encompassing a "whopping" 1,100 square feet, these were considered to be nice homes, and in comparison with some of the small shack-style houses in Atchee, this was true. All of the company-owned houses in Atchee were identified by their residents and listed in company records by job titles. In 1919, the four houses were titled, from north to south, as an "employee residence," "car foreman's house," "boilermaker's house," and "roadmaster's house." The two views on this page are of the "roadmaster's house," as seen in the late 1920's—at which time it was known as the "machinist's house." The railway's head machinist, Robert Hester, was living there (with his family). This was the last of the frame company houses at the south end of the boardwalk.

ROBERT C. BOULTER COLLECTION

THE BACKSIDE of machinist Robert Hester's house during a winter in the 1920's. The back-porch addition originally was much smaller, but was rebuilt to almost the entire width of the building in the early 1920's. It is unclear whether or not the additions were paid for by the railway or the employees, but most likely the construction was done at the employees' expense. A coal box sat in the foreground, while in the background, just in front of the house, caboose No. 3 was resting on the mainline. As they only lived a few hundred feet away from the shops and waiting engines, the employees nearly always were ready to be called out. There was no communication other than by a knock on the door late at night or early during the morning, which meant one was needed to be ready for work in the next few minutes.

ROBERT C. BOULTER COLLECTION

YOUNG WILLIAM (BILL) HESTER stood on the snowy road through Atchee during the middle of the winter. Unfortunately, Bill was killed when he fell from a boxcar in the Colorado & Southern Railway yard in Denver during 1917. This photograph was taken as the camera was pointed southwest, toward two of the other employee's houses, located near the end of the boardwalk—close to "School House Canyon." The building at the left actually was one of the older passenger cars of the Uintah Railway, after it was taken out of service. This coach body sat in this spot for several years and then disappeared, its fate unknown. With a few windows boarded up, including the clerestory windows, and a porch installed on the far side, out-of-service passenger cars were good for helping the housing shortage.

ence. Being the only non-company-owned entity in Atchee would have its advantages over the years, in effect causing the Lutons to become very important people in the politics of such a small town. They started their grocery store in a small building next to the track, across from the water tank. That building would be added onto over the years, time after time, until they had a combination grocery, hardware store, meat market and post office. There was always time to do something more, especially for Ruby, who would become involved in almost every aspect of what went on in Atchee. In the true spirit of American free enterprise, the Lutons were the only people in town who were not usually direct employees of the railway, and in turn, they ended up with as many responsibilities as could be doled out. Metters ran the store. Ruby became the post mistress, ran the Uintah Railway's company hotel, filled-in as secretary at the company shops, ran the voting system for Garfield County and was the unofficial "doctor" for most of the townspeople's minor ailments.

RODGER POLLEY COLLECTION

SOME SMALL two-room apartments were constructed in Atchee to help with overcrowding. These were simple frame structures, with board-and-batten siding and tin roofs.

RODGER POLLEY COLLECTION

BEHIND THE HOTEL, **three other frame buildings were built by the Uintah in 1904. These houses were not designed with the thought of families living in Atchee, as were the cottages on the other side of the hotel. Instead, these houses were built to the section house design, with a single A-frame roof from the front to the back. As Atchee grew rapidly, the number of employees needed to keep the railway's town running created the need for more housing. Along with building several other small residences in town, the Uintah's management added onto these buildings over the years. By 1911, the first house and the last house had been expanded, with the last house nearly doubling in floor space. The front of this larger house was seen here as it looked around 1911. This building was known as the "car foreman's house" and was used as a boarding house for laborers. This building was especially needed with the car-building projects in 1910–1912, and during the extension of the railway to Rainbow.**

The store itself was an assortment of small lean-tos and add-ons that amounted to a maze of small rooms. The building was made up of a large area in the main building, used strictly for merchandise, a separate area for meat, a large stockroom, the post office room, a small room for a pool hall, and a small residence where the Lutons lived. Mostly dried and canned goods filled the shelves at the Luton's store. Early on, shipments were made from Mack to Atchee on the train to fill the shelves, but by the late 1920's, Metters Luton would make the trip to Mack, Fruita, and Grand Junction almost every week, usually on Fridays. It was then that small, individual lists of goods was given to him by the townspeople so they could be sure that Luton got

ROBERT C. BOULTER COLLECTION

FROM THE SOUTHEAST **end of the boardwalk, you are looking northwest, toward the railway yard. These scenes were photographed from the same location during different seasons during the 1930's. In this view, notice the addition that was built on the second employee's house, at the left. This was an old addition dating back to around 1911. The main boardwalk turned left toward the rest of the residential area of Atchee, at the far end, and to the right, toward the enginehouse and shop office. The buildings on the right were built in the late teens, as the Uintah reached its peak of activity. More housing was needed, and a whole group of small houses was built between the boardwalk and the car shops, in the northeast. The long porches were added later, and another row of poplar trees was planted on that side of the walkway. The boardwalks were kept lighted in the evening, and the electric pole in the foreground shows an early street lamp hanging over the walkway.**

THE COMPANY HOUSES, **or "cottages" as they were known, were all quite modern and well equipped at the time they were built. The Uintah's management knew even before it was constructed that the only way that they could keep good workers was not only to pay good wages, but also to provide the basics of life in such an inhospitable area. Running water, sewer systems and steam heat were installed when the houses and Uintah Hotel were built. Supplied from the main water tank, the houses had nearly as much water as they wanted, with pretty good pressure. The steam came from the boiler plant at the machine shop, and it supplied the houses with as much heat as they wanted in the winter.**

JIM McCOY COLLECTION

ROY McCOY PAUSED on his way to work on the boardwalk in front of his house. He was carrying his lunch in a metal pail, to be eaten in the cab of the engine somewhere on the other side of Baxter Pass. In the background is stock car No. 401, sitting on the siding in the yard. Notice the piece of Uintah freight car that was being used as a fence post. The "Capacity 40,000" notation referred to the amount of weight a car could carry. Since spare lumber was scarce, this piece of one of the Uintah's cars probably came from a wreck where parts were salvaged and brought back to Atchee.

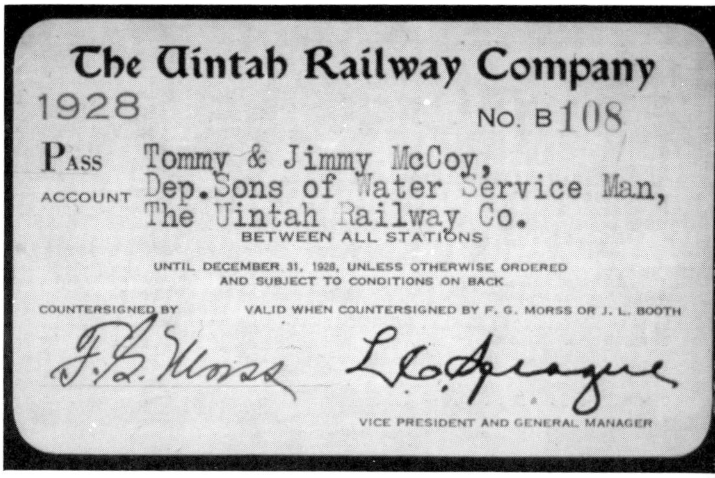
The Uintah Railway Company
1928 No. B108
Pass Tommy & Jimmy McCoy,
Account Dep.Sons of Water Service Man,
The Uintah Railway Co.
BETWEEN ALL STATIONS
UNTIL DECEMBER 31, 1928, UNLESS OTHERWISE ORDERED AND SUBJECT TO CONDITIONS ON BACK
COUNTERSIGNED BY VALID WHEN COUNTERSIGNED BY F. G. MORSS OR J. L. BOOTH
F.G. Morss L.C. Sprague
VICE PRESIDENT AND GENERAL MANAGER

RICHARD A. RONZIO COLLECTION

what they needed. The Luton's prices were relatively high, compared with those of the stores in Fruita, and many families bought most of their goods there, having them shipped up to Atchee on the Uintah, rather than pay through the nose at Luton's. Still, there were necessities that residents would forget, or desperately needed items had to be purchased at a premium price at the Luton store. Friday afternoons in the 1930's would see Mr. Luton arriving from the Grand Valley in his truck, and thereupon,

AS WITH ALL SMALL TOWNS across the United States, the national pastime was baseball, and Atchee was no exception. Teams from Mack, Atchee and Dragon would play one another and go to other towns in the region on the weekends to "play ball." The McCoy family was seen next to their house during the 1930's. In the center was Jim, and to the left was his older brother, Tim, while their father, Roy McCoy, was to the right—and they were all dressed up in homemade sweaters and caps, featuring the company logo. By this time, organized baseball had become a more regional sport, and men from all over the railway line were part of one team, traveling on weekends by way of automobile to Grand Junction and other locations to play. This became a real object of pride for the employees, and in many cases, a prospective new employee's pitching or hitting ability was every bit as important as his train experience. After all, there were plenty of out-of-work trainmen to choose from during the 1930's, so they might as well also be able to play good baseball.

JIM McCOY COLLECTION

JIM McCOY COLLECTION

IN SUCH AN ISOLATED community, horse power—or in this case, burro power—was still needed from time to time for odd jobs. There were no farm tractors or big trucks in Atchee, and burros were still used on occasion to move things around town. Their usefulness decreased dramatically after automobiles came to town during the 1920's. By the time this picture of Roy McCoy and his sons was taken in the 1930's, the little burros were used for fun as much as for work. This photograph shows the guys on the boardwalk in front of the houses, next to the mainline.

he would unload the supplies for the next week. Special items that had been requested were quickly gathered up by the ladies in town, as well as items like fresh fruit, fresh meats, milk and sanitary items that might have been forgotten on the lists.

Driving his truck on the dirt road between Atchee and the Mesa County roads near the farms in Mack would occupy a large part of Mr. Luton's life, as vehicle travel increased through the 1920's and 1930's. Not only did Mr. Luton travel over the roads almost every week, but he became the foremost advocate in Atchee to have the roads maintained by Garfield County. His constant hounding of the county authorities eventually caused them to try and solve the issue; however, as Mr. Luton later found out, it was to the county's advantage. Their solution was to purchase a road grader to help fill the deep ruts in the road, which appeared after every rain storm, or during the spring thaw. The job of grading the road was then passed on to Mr. Luton, who obtained the contract to maintain the road. The catch was that the contract only paid a small amount per month, not nearly enough to compensate Mr. Luton for more than a few hours of work.

JIM McCOY COLLECTION

YOUNG JIM McCOY was standing in front of a string of boxcars in the yards at Atchee dressed up as a cowboy. In the background, over the top of the boxcars, is the roof of the ice house. Notice the "diamond" placard on the door of the boxcar, which often signified gasoline, explosives or other dangerous freight needed to be handled with special care.

ROY ENO JR. COLLECTION

THIS ATCHEE SCENE was photographed from the south, looking north toward the machine shop and enginehouse. Although there were seven of the larger frame employee houses, most of the employees lived in smaller, more humble houses like these. The white house in the foreground was owned by long-time employee Roy Eno. While many of the newer employees tried to work their way up the seniority ladder to get into one of the larger homes, Mr. Eno and his family lived here and made it home, not wanting to move out when the opportunity arose. At the lower left of the scene, the small mound of dirt, with the stovepipe sticking up, was the location of the family's root cellar. Many families utilized small root cellars to store canned goods and supplies, and there were as many, or more, root cellars in town as there were automobile garages.

This shifted the burden from the county government and effectively made Mr. Luton the scapegoat for why the roads were bad. He had the contract and the grader, so why were the roads not in tip-top shape at all times?

One of the final money-making ventures the Lutons added to their store was a Texaco filling station. A gravity-fed pump and few new signs on the front of the store during the late 1930's allowed people to "gas up" at Atchee before heading out for Mack. "Gasing up" had always been a problem, in that you had to make sure your automobile was filled to the brim with gasoline before leaving Mack for Atchee, otherwise you might not make it back to Mack again. That problem was amplified on Sundays, when most of the railroaders who had gone into the Grand Valley could not find any filling station open, so they could buy gasoline.

ROY ENO JR. COLLECTION

A CONSTANT SOURCE of meat was provided by the large herds of mule deer, which roamed the valleys around Atchee. Before the days of game management and over-hunting, families in Atchee could expect to hunt deer all year 'round. Many deer roamed the numerous valleys between Carbonera, Colorado, and Dragon, Utah, leaving ample opportunity for the Uintah's employees to shoot as much meat as could be used. This view of Roy Eno Jr. and his father, with a four-point buck, was taken in front of their house in Atchee.

The Lutons also were the natural choice for the people to run the post office. The Uintah Railway frowned on having their employees working for the U.S. Post Office Department, and most of the wives in town had families to raise, with no help from dad during the day. When the Lutons arrived in town, it was only a few years before they had the postal contract. The previous post office had been located at Rogers' ranch, located a quarter-mile up the grade on the other side of West Salt Wash Creek. This proved to be unpopular with the townsfolk, as they had to walk to the ranch everyday to get their mail. The Uintah also disliked this arrangement, as the public had to cross their Bridge 28-A to get to the mail.

ROBERT C. BOULTER COLLECTION

THE PRIMITIVE BALL FIELD where Atchee teams originally played games was in the large open space between the hotel and the south leg of the wye. This was changed to the east side of the wye after the car shops were built, ca. 1910. It was in this new location where the summertime games were played. The thick sagebrush was cut from the area, and the natural wild grass formed the base of the field. Preparation of the ball field had to be done several times a year, as the sagebrush and other bushes would quickly sprout up from their roots and fill the field with knee-high vegetation. The surface was not perfect, either, as the field was far from level. Home plate was up near the wye, with first base placed out to the east toward the mountain behind town, and third base to the left in the direction of West Salt Wash. Basically, this left the field canted, as first base was several feet higher than third, thus giving a whole new meaning to "home field advantage." This view shows most of the community had turned out to watch one of the baseball games. The "grandstand" was built by the shop men to accommodate everyone, and the board seats filled with employees and their families, watching their co-workers take on the out-of-towners. In the foreground was the track that led up the coal dump for the boilerhouse. Notice the large post near the switchstand, with cables tied off to support the tall smokestack on the boilerhouse.

ENGINEER ROY ENO made his way from his home at the west end of Atchee toward the ball field on the east side of the wye, about 1920. Uniforms were generally used from wherever else employees had played, and sometimes, the field was filled with the names of other towns and companies, whose uniforms were now being used by the Atchee team. Mr. Eno was dressed in a full-length coat, as the chilly spring air in Atchee Canyon could be somewhat cooler than the places where the visiting teams came from.

ROY ENO JR. COLLECTION

COLORADO RAILROAD MUSEUM COLLECTION

AS THE TOWN of Atchee was being dismantled, several of the old Pullman sleeping cars that had been used for housing were pulled up and sold to individuals in the Grand Valley. This car sat near the schoolhouse and was used as a residence for several years, but it is not known what its original car number was. A cast-iron cook stove had been pulled out of the car. The rear platform had been enclosed with a screen. Observe the railway grade as it climbed Baxter Pass on the distant snow-covered mountainside.

HERE IS A VIEW of the small residences in Atchee in winter, framed by Baxter Pass in the background. Many times, the wives in town would hold an afternoon get-together in one of the houses at this end of town, and after seeing smoke from the oncoming train on Baxter Pass, they would know that in about an hour, the men in the crew would be back in Atchee.

ROBERT BOULTER COLLECTION

RODGER POLLEY COLLECTION

ONE RESIDENCE used for company employees was built on the east side of the wye, near the bridge. It is unclear how long it stood, but it was not around for long. Here, a couple of employees ham it up for the camera, pretending to crank-start an old automobile (minus a motor). The railings for the Uintah's Bridge 28-A were in the background.

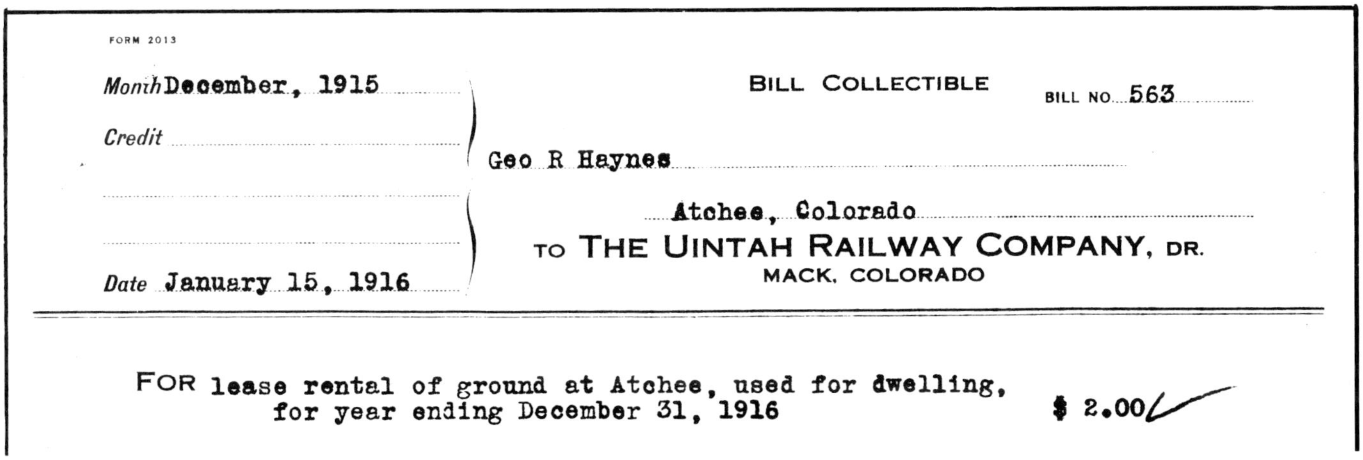

FORM 2013

Month December, 1915

Credit

Date January 15, 1916

BILL COLLECTIBLE

BILL NO. 563

Geo R Haynes

Atchee, Colorado

TO THE UINTAH RAILWAY COMPANY, DR.

MACK, COLORADO

FOR lease rental of ground at Atchee, used for dwelling, for year ending December 31, 1916 $ 2.00

KELLEY BARLOW COLLECTION

A CLOSE-UP VIEW is shown here of the front of the house, pictured at the left. These small houses usually were cramped for space, but most of the people who lived in them made the best of it. Porches and lean-tos were often added over time to help alleviate summer heat and provide storage space.

RICHARD A. RONZIO COLLECTION

THIS VIEW OF ATCHEE, taken from east of town on the hill above the wye, shows it ca. 1930. The buildings that made up the car-shop complex had been expanded to their largest size by this time, with various outbuildings and storage sheds making up the large group of roof tops in the middle of the picture. The boilerhouse coal tipple and its approach can clearly be seen. The creation of a coal tipple at the boilerhouse, and the way the cars hauling coal were switched to that track, greatly reduced the amount of storage trackage east of the primary coal tipple. This was the main reason why the storage track on the opposite side of Bridge 28-A had to be constructed.

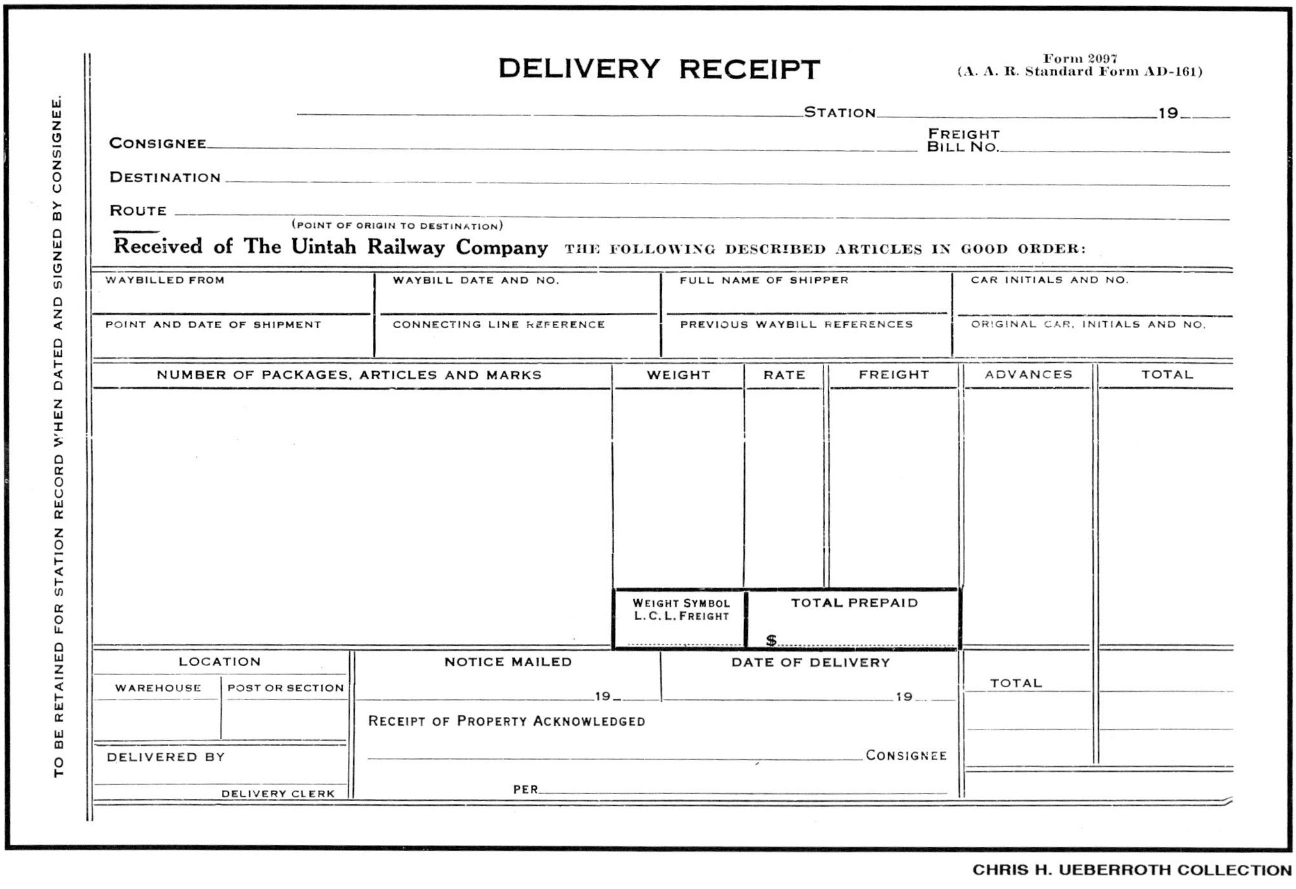

TO BE RETAINED FOR STATION RECORD WHEN DATED AND SIGNED BY CONSIGNEE.

DELIVERY RECEIPT

Form 2097
(A. A. R. Standard Form AD-161)

______ STATION ______ 19___

CONSIGNEE ______ FREIGHT BILL NO. ______

DESTINATION ______

ROUTE ______
(POINT OF ORIGIN TO DESTINATION)

Received of The Uintah Railway Company THE FOLLOWING DESCRIBED ARTICLES IN GOOD ORDER:

WAYBILLED FROM	WAYBILL DATE AND NO.	FULL NAME OF SHIPPER	CAR INITIALS AND NO.
POINT AND DATE OF SHIPMENT	CONNECTING LINE REFERENCE	PREVIOUS WAYBILL REFERENCES	ORIGINAL CAR, INITIALS AND NO.

NUMBER OF PACKAGES, ARTICLES AND MARKS	WEIGHT	RATE	FREIGHT	ADVANCES	TOTAL
	WEIGHT SYMBOL L. C. L. FREIGHT	TOTAL PREPAID $			

LOCATION — WAREHOUSE | POST OR SECTION

NOTICE MAILED ______ 19___

DATE OF DELIVERY ______ 19___

TOTAL

RECEIPT OF PROPERTY ACKNOWLEDGED

DELIVERED BY ______ CONSIGNEE

DELIVERY CLERK ______ PER ______

CHRIS H. UEBERROTH COLLECTION

THIS AGREEMENT, Made this First day of January 1916,

between THE UINTAH RAILWAY COMPANY as first party, and George Haynes

as second party,

WITNESSETH, that the first party hereby demises and leases to the second party a certain piece or parcel of land, situate at Atchee, Colorado,

described ~~as follows~~: on typewritten sheet marked 28-7 a-5 on

a plat of said premises ~~being~~ hereto attached, and made a part hereof.

TO HOLD the same unto the second party as tenant at will of the first party, and for the rent of Five — Dollars ($) per annum, from the First day of January A. D. 1916, payable in advance.

Said first party hereby reserves the right to terminate this agreement and said tenancy, and to take possession of and re-enter upon said premises at any time hereafter, after having given to the second party, or left upon said premises ten (10) days' notice, in writing, of its intention so to do, and requiring said party to quit and deliver up possession of said premises, and in the event that this agreement is terminated by notice as hereinbefore provided to be given by said first party, before the expiration of the term for which rent has been paid in advance, then and in such event the first party shall refund to the second party the unearned rent money so paid in advance and said second party shall also have the right to quit and deliver up possession of said premises, and terminate the tenancy hereunder, by giving to the said first party, or one of its duly authorized agents ten (10) days' notice, in writing, of such intention, but in such event said second party shall not be entitled to a refund of any part of the rent money which may have been paid in advance.

Said second party hereby agrees to become and continue to be the tenant at will of the first party, under the conditions hereof, to pay said rent, as above provided, and to keep said premises in good order and condition, and to quit and deliver up possession of said premises to said first party upon the expiration of the time mentioned in any notice to quit, given as aforesaid; and that upon the expiration of such time, the said tenancy shall cease, and said first party shall have the right forthwith to re-enter upon and occupy said premises, without any further or other notice, proceeding or process whatsoever.

The said second party hereby agrees not to use or occupy said premises for any other purpose than

[illegible]

and not to assign, or transfer, this agreement, nor underlet said premises, without the written consent of the first party being first obtained.

The said second party hereby agrees to pay all taxes and assessments levied on all improvements erected on said premises by said second party; and also agrees to waive all claims and demands against the first party for damages on account of injury to said improvements, and appurtenances thereto and the contents thereof, and to indemnify, protect and save harmless the first party, against all claims and demands for such damages, caused by said first party or any of its employees. The lessee specially assumes all risks and agrees to indemnify, protect, and save harmless said lessor, from all claims of whatsoever character for damages resulting to the property of said lessee by reason of fire originating from the engines, or locomotives, of the lessor or otherwise, and resulting in the burning or destruction of, or injury to, the property of the lessee.

That said second party shall have the right to remove said improvements at the termination of said tenancy, if not in default as to any of the covenants herein mentioned to be performed by said second party; said removal to be made before the expiration of the time mentioned in any notice given for the termination of this agreement as herein provided.

IN WITNESS WHEREOF, The parties hereto have executed this agreement, in duplicate, the day and year first above written.

THE UINTAH RAILWAY COMPANY,

WITNESS:

By [illegible signature]

General Manager.

George Haynes (Seal)

(Seal)

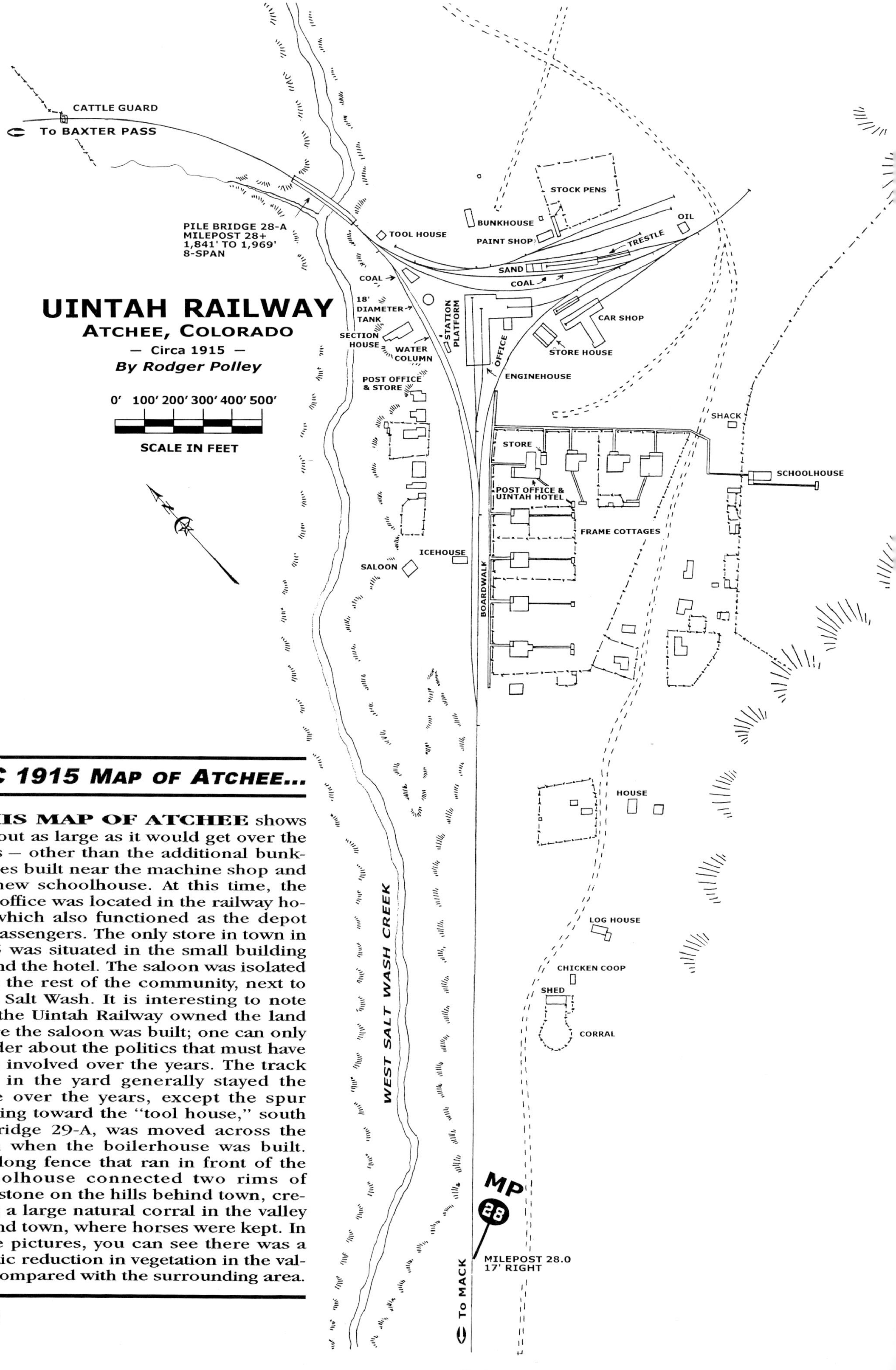

ICC 1915 Map of Atchee...

This map of Atchee shows it about as large as it would get over the years – other than the additional bunkhouses built near the machine shop and the new schoolhouse. At this time, the post office was located in the railway hotel, which also functioned as the depot for passengers. The only store in town in 1915 was situated in the small building behind the hotel. The saloon was isolated from the rest of the community, next to West Salt Wash. It is interesting to note that the Uintah Railway owned the land where the saloon was built; one can only wonder about the politics that must have been involved over the years. The track plan in the yard generally stayed the same over the years, except the spur heading toward the "tool house," south of Bridge 29-A, was moved across the wash when the boilerhouse was built. The long fence that ran in front of the schoolhouse connected two rims of sandstone on the hills behind town, creating a large natural corral in the valley behind town, where horses were kept. In some pictures, you can see there was a drastic reduction in vegetation in the valley, compared with the surrounding area.

3-07-99
25

*5200
*5200
*403
*5603
1 *
-5082

m Domestic Straight Bill of Lading, adopted by Carriers in Official, Southern and Western Classification territories, March 15, 1922, as amended August 1, 1930.)

THIS SHIPPING ORDER must be legibly filled in, in Ink, in Indelible Pencil, or in Carbon, and retained by the Agent.

Shipper's No.______________
Agent's No.______________

THE UINTAH RAILWAY COMPANY

ECEIVE, subject to the classifications and tariffs in effect on the date of the issue of this Shipping Order.

______________________________ 193___

n ______________________________

perty described below, in apparent good order, except as noted (contents and condition of contents of packages unknown), marked, ed, and destined as indicated below, which said company (the word company being understood throughout this contract as mean- person or corporation in possession of the property under the contract) agrees to carry to its usual place of delivery at said tion, if on its own road or its own water line, otherwise to deliver to another carrier on the route to said destination. It is y agreed, as to each carrier of all or any of said property over all or any portion of said route to destination, and as to each t any time interested in all or any of said property, that every service to be performed hereunder shall be subject to all the con- not prohibited by law, whether printed or written, herein contained, including the conditions on back hereof, which are hereby to by the shipper and accepted for himself and his assigns.

(Mail or street address of consignee—For purposes of notification only.)

ed to______________________________

tion______________ State of______________ County of______________

Route______________________________

Delivering Carrier______________ Car Initial______________ Car No.______________

No. Packages	DESCRIPTION OF ARTICLES, SPECIAL MARKS, AND EXCEPTIONS	*WEIGHT (Subject to Correction)	Class or Rate	Check Column

Subject to Section 7 of conditions, if this shipment is to be delivered to the consignee without recourse on the consignor, the consignor shall sign the following statement:

The carrier shall not make delivery of this shipment without payment of freight and all other lawful charges.

....
(Signature of consignor.)

If charges are to be prepaid, write or stamp here, "To be Prepaid."

....

Received $..............................
to apply in prepayment of the charges on the property described hereon.

....
(Agent or Cashier)

Per..............................
(The signature here acknowledges only the amount prepaid.)

Charges advanced:

$..............................

★ If the shipment moves between two ports by a carrier by water, the law requires that the bill of lading shall state whether it is "carrier's or shipper's weight."

Note—Where the rate is dependent on value, shippers are required to state specifically in writing the agreed or declared value of the property.

The agreed or declared value of the property is hereby specifically stated by the shipper to be not exceeding

______________________________ per ______________________________

______________________________Shipper. ______________________________Agent.

Per______________________________ Per______________________________

Permanent post-office address of shipper ______________________________

10927 J. J. COLLINS' SONS, INC., PRINTERS, CHICAGO

CHRIS H. UEBERROTH COLLECTION

CHRIS H. UEBERROTH COLLECTION

UINTAH ENGINE No. 20 was ready to be shipped from the Baldwin Locomotive Works during June of 1905. This 0-6-2T originally was fitted with two sand domes; however, she was rebuilt with only one sand dome and an electric generator. This engine had a 125-ton rating between Mack and Carbonera, whereas she had only a 35-ton rating between Atchee and Baxter Pass.

UINTAH 0-6-2T TANK ENGINES...

DAILY MAINTENANCE was part of every trip over Baxter Pass. As with any railway, mechanical failures with the brakes, air lines or valve gear were possible and could lead to injury or death. On the Uintah Railway, going up and down Baxter Pass was not just some remote chance of an accident, but an almost certain recipe for disaster if maintenance on these systems was not done perfectly. Here, No. 20 was sitting in the yard at Atchee, as the engineer checked and oiled the rigging under the frame of this 0-6-2T.

RICHARD A. RONZIO COLLECTION

JOHN AND HAZEL CARSON COLLECTION

THE REGULAR PASSENGER TRAIN, with its usual 0-6-2T tank engine and single passenger car, was being helped by Shay No. 2, seen here crossing Bridge 28-A, heading for Baxter Pass. Trains going up the next six miles of constant 7.5-percent grade could use all the power they could muster. Helper engines normally were needed on this severe grade.

LOCOMOTIVE No. 20 was ready for departure, with two passenger coaches and a Shay locomotive under steam (at the rear) to make the climb up Baxter Pass ca. 1910. Notice the block-style lettering on the side of the water jacket. The fireman or engineer was leaning against the water stand pipe, which rested on planks set in cinders at its base.

DAN ABBOTT COLLECTION

RICHARD H. KINDIG COLLECTION

UINTAH TANK ENGINE No. 20 was next to the water column at Atchee. This was the regular passenger train for most days, with a car half full of passengers and nearly as many parcels and other items to be delivered at various points along the line, including groceries. Although the letters on the side of the combination car are out of focus and illegible, they read: "AMERICAN RAILWAY EXPRESS AGENCY."

DR. LESTER WILLIAMS COLLECTION

UINTAH ENGINES AT ATCHEE...

THE INSCRIPTION with this photograph read: *Just out of shops near Atchee depot. John Simpson in the cab. Left to right on the ground, Glen Griffin, Robert Hester, Scott Shafer and Ben Waddell. Engine painted olive green, with yellow and red stripes. Henry Smith was engineer.* The two small tank engines were rebuilt and modified several times over their 35 years with the railway and this was No. 20's final overhaul at the shop in Atchee during the mid-1920's. The coal bunker had been added onto three times, as noted by the three major seams on the back of the engine. As originally built, the coal capacity was far from sufficient for the longer runs over the line, especially after the extension of the line to Watson and Rainbow in 1911. The original coal bunker was added to later with the next section of the bunker walls being extended up about two feet. During this last rebuilding job, it was extended one more time, to the top of the cab. Notice the back-up light at the rear of the coal bunker.

BALDWIN 0-6-2T SIDE-TANK engine No. 20 was commonly used over the Uintah's steep grade between Atchee and Baxter Pass. Here, this 55-ton engine had paused in Atchee before taking another train up the line. She was able to haul one coach (35 tons) over the seven-mile line to the summit—which had grades running from 4.0 percent to 7.5 percent—averaging 55 minutes for the upgrade run. The cylinders measured 13x18 inches, while the drivers were of 34-inch diameter, and the total weight was 68,670 pounds. In this scene, the newly reworked and shiny locomotive was sitting on the mainline in front of Luton's store.

ALLAN C. LEWIS COLLECTION

U.RY
20

20
U.RY
20

ROBERT C. BOULTER COLLECTION

THE WATER TANK AT ATCHEE was a required stop for all engines before leaving in either direction. The 28 miles through the desert to Mack was too great a distance to risk going without taking on water. It was better to be safe than sorry on the steep climb up Baxter Pass. The front pilot of Engine 21 was in disrepair, as many of the wooden staves had been knocked out during coupling. This early picture shows the engine before she was rebuilt with a larger coal bunker and only one sand dome. No. 21 and her sister engine, No. 20, were used primarily as passenger engines over Baxter Pass, pulling the customary one car. Here, she was pulling combination car No. 1, as the train crew stood still for the picture. The enginehouse was in the distant background, and the edge of the small coal bunker was visible at the far left.

CLIFF HAYCOCK COLLECTION

THE MAINLINE AND SIDING in front of the company houses were one of the busiest portions of the yard at Atchee. One curve of the wye connected in front of the hotel, as seen here. No. 21 was backing up at the wye.

ENGINE No. 21 sat behind one of the Shays in this picture, photographed on September 3, 1934, from on top of the trestle that led up to the top of the coal chute. The coal bunker on the back of the engine shows the three distinct additions to its capacity, as this 0-6-2T was modified over the years, with the final addition matching the curvature of the cab roof. Constant stops at coaling facilities became a hassle early in her use as a passenger engine, and the need for more capacity outweighed any superficial desire for looks or "balance." Notice the electric backup light added to the engine during her last rebuilding job.

OTTO C. PERRY PHOTO – DENVER PUBLIC LIBRARY, WESTERN HISTORY DEPARTMENT

ssenger Engine
ntah Ry. Co.
tchee, Colo.

21
URY
21

ROBERT C. BOULTER COLLECTION

FIREBOX ASHES and clinkers from the Uintah's engines were dropped near the enginehouse, and (in turn) shoveled over beside the water tank. Here, Engine 21 and Caboose 1 were getting ready to depart from Atchee, and the engineer was standing in the doorway of the cab with his oil can. An extra air-reservoir tank was located in front of the bell, between the sand and steam domes.

RODGER POLLEY COLLECTION

PASSENGER ENGINE No. 21, was near the boilerhouse, which had board-and-batten siding. In this picture, the locomotive had been recently overhauled and had a fresh paint job. Riveted patches on the water jackets are evidence of the machine shop's continuing fight against rust. The bunker had been heaped high with coal for the next assault on Baxter Pass.

RODGER POLLEY COLLECTION

PARCELS AND LUGGAGE were unloaded from the combination express-mail coach coupled to No. 21. A number of wheelbarrows and dollies were being used to distribute loads around Atchee. The air-reservoir tank mounted in front of the cab had a different position than the one on Engine 20.

DAN PRICE COLLECTION

THE STATION SIGN on the west side of the building reads: "ATCHEE, Elev. 6340, Mack 28.3 M." The passenger platform was just a few feet away from the enginehouse, and this was the best place to hang the station sign, even though there was no official depot provided here. Three young men from Atchee stood on top of one of the two different kinds of handcars, sitting next to the building. They probably thought it was a real honor to get their picture taken on top of the handcar, as they were off limits to all children. This was especially true after one of the cars came up missing during the early 1920's, only to be found several miles downgrade toward Carbonera. Several children were disciplined for having been seen playing next to it earlier in the day. These small push cars were equipped with a simple brake rigging that allowed them to descend Baxter Pass on their own. You can see the small brake levers sticking up on the sides. Although official company policy did not allow for the use of the small "speeders," they often were attached behind a freight train and pulled to the top of the pass. Once there, townsfolk from Atchee would get off the train, unhook the car and (by using the brake) descend Baxter Pass. Section men and employees needing a ride from the top down often used these cars. Courting couples also borrowed the little hand cars, as there was nothing like getting a good scare out of one's date coming down the 7.5-percent grade at a fast pace. In the autumn, mothers and daughters were taken to the top of the pass and allowed to use the cars to pick choke cherries, a wild bittersweet fruit that grew all along the right-of-way from Shale Tank to below Morro Castle. Large buckets were taken along and filled by the ladies who handpicked the fruit a short distance off the tracks.

RICHARD A. RONZIO COLLECTION

IT IS NOT KNOWN whether or not railcar No. 50 was used on a regular basis by the doctor during the teens or if this was some kind of special trip or occasion for a group of health care workers. The red cross on the placard indicates the special purpose of the this trip. The Uintah had a doctor on staff who was stationed in Dragon, Utah. Regular weekly runs over the entire line allowed him to visit all of the towns along the line and provide basic care for the varying ailments of the townspeople. Most of the time the doctor rode on the passenger train to do his rounds, and would stay at the hotels in Atchee and Mack overnight before returning home to Dragon. The parent company of the Uintah, Gilson Asphaltum Company, saw that the doctor was located in Dragon not only for the railroaders' benefit, but also for the miners at Dragon, Country Boy, Rector and Rainbow. Although the railroaders had their share of accidents and sickness, the many miners concentrated in the Dragon area required many more trips to the doctor.

Uintah Railcar No. 50...

DAN ABBOTT COLLECTION

THE MAN INSPECTING the front wheels of the Mack Brothers' Railcar 50 was getting plenty of help from the town's dogs. This photograph was taken near the back of the machine shop, on the wye trackage. Under the feet of the three men on the right, the small side track running off to the right ran to a small storage shed, where a handcar was stored.

JIM McCOY COLLECTION

A GROUP OF EMPLOYEES, including company claim agent and photographer Frank Kennedy (standing at the left), posed for a photograph in front of Railcar 50 after getting back from a trip to Dragon, Utah. The motorman of the vehicle had picked a bunch of flowers on top of the pass for the lucky lady. At this time, the front headlight on the railbus was an electric carbide lamp, and the canister can be seen hanging off the left front of the vehicle. Between the rails in the foreground was a small ash pit where the engines dumped their ashes. This ash pit was filled-in later, and ashes were dumped behind the enginehouse. This pit had to be cleaned by hand, and on a regular basis, an employee would spend most of the day shoveling cinders and ashes into a wheelbarrow. Then, he took it to the edge of West Salt Wash, one load at a time. This is an older image of the enginehouse doors on the southeast stall. The doors simply swung shut from both sides. Later, the doors were rebuilt, and the one seen here had an additional folding part added to it and more cross braces placed on the front and back.

F. A. Kennedy, Claim Ag't,
Uintah Ry. Co.,
MACK, COLO

RODGER POLLEY COLLECTION

PETS ALWAYS WERE AROUND Atchee—especially dogs. Here, Metters Luton (standing) and Scott Shafer were petting a local dog in the yard of the Uintah Hotel. In the background was the warehouse for the Uintah's car shop, built on concrete piers to keep the lumber away from the damp earth, while at the same time, it did not require the expense of a full foundation.

ONE-WAY FARES.

Side Line Number.	Head Line Letter. / BETWEEN	A Mack, Colo.	B Salt Wash, Colo.	C Carbonera, Colo.	D Atchee, Colo.	E Shale, Colo.	F Baxter Pass, Colo.	G Deer Run, Colo.	H Columbine, Colo.	I McAndrews, Colo.	J Wendella, Colo.	K Sewall, Colo.	L East Vac, Colo.	M Coluta, Colo.	N Dragon Junction, Utah.	O Ute, Utah.	P Watson, Utah.	Q American, Utah.
1	Mack, Colo.	$ ---																
2	SaltWash, Colo.	0.80	$ ---															
3	Carbonera, Colo.	1.40	0.60	$ ---														
4	Atchee, Colo.	2.00	1.20	0.60	$ ---													
5	Shale, Colo.	2.25	1.45	.85	0.25	$ ---												
6	Baxter Pass, Col.	2.40	1.60	1.00	.40	†0.15	$ ---											
7	Deer Run, Colo.	2.55	1.75	1.15	.55	.30	†0.15	$ ---										
8	Columbine, Col.	†2.65	1.80	1.25	.65	.40	.25	†0.15	$ ---									
9	McAndrews, Col	2.80	2.00	1.40	.80	.55	.40	.25	0.25	$ ---								
10	Wendella, Colo.	2.90	2.10	1.50	.85	.60	.45	.35	.25	†0.15	$ ---							
11	Sewall, Colo.	2.95	2.15	†1.55	.95	.70	.55	.40	.30	.25	†0.15	$ ---						
12	East Vac, Colo.	3.30	2.50	1.90	1.30	1.05	.90	.75	.70	.50	.45	0.35	$ ---					
13	Coluta, Colo.	3.50	2.75	2.15	1.55	1.30	1.15	1.05	.90	.75	.65	.60	*0.30	$ ---				
14	Dragon Jct., Ut.	3.50	2.95	2.35	1.75	1.50	1.35	1.25	1.10	.95	.85	.80	* .45	0.25	$ ---			
15	Ute, Utah	3.75	3.30	2.70	2.15	1.90	1.75	1.60	1.50	1.30	1.25	1.15	.80	.55	0.40	$ ---		
16	Watson, Utah	4.00	*3.60	*3.00	*2.40	*2.15	*2.00	*1.85	*1.75	*1.60	*1.55	*1.45	*1.10	* .85	.50	*0.30	$ ---	
17	American, Utah	4.10	*3.70	*3.10	*2.50	*2.25	*2.10	*1.95	*1.85	*1.70	*1.65	*1.55	*1.20	* .95	.60	* .40	†0.15	$ ---
18	Rainbow, Utah	4.15	*3.75	*3.15	*2.55	*2.30	*2.15	*2.00	*1.90	*1.75	*1.70	*1.60	*1.25	*1.00	.65	* .45	.25	*0.20
	DavisCanon, Col	No	rates	in eff	ect.													
	Temple, Utah	No	rates	in eff	ect.													

† Decrease. * Increase.

TRAIN FARES.

(a) From stations where a ticket office is not maintained, conductors will collect ticket fares shown above.
(b) From stations where a ticket office is maintained, conductors will collect fares as follows:—

Where the Adult Ticket Fare Shown Herein is	Collect the Amount Shown Below in Addition to the Ticket Fare.
50 cents or less	10 cents.
More than 50 cents and not more than $1.00	15 cents.
More than $1.00 and not more than $1.50	20 cents.
More than $1.50	25 cents.

(c) For each child five years of age and under twelve, one-half of the total charge provided for in paragraph (b) will apply, sufficient to be added when necessary to make fare end in 0 or 5.

Where cash fare is paid as a part of a continuous trip in connection with a ticket or a mileage book, the ticket fare may be accepted.

[3]

CHRIS H. UEBBEROTH COLLECTION

ED AND NANCY BATHKE COLLECTION

AT THE TIME THIS PICTURE was taken of the front of Uintah Railway's engine No. 10, around 1909, she was the pride and joy of the railway's small fleet of seven locomotives. Sitting on the pilot of this Consolidation, with his foot resting on the coupler, was George Lyman. He was the main engineer assigned to the run from Mack to Atchee, and No. 10 usually had the role of taking the daily passenger train across the desert. The desert run for the passenger train was the most sought after engineer position at the time, and Engineer Lyman and this 2-8-0 performed the job for years without any problems—until the wreck took place in 1911 at Bridge 1-A. This photograph shows the scene as you look at the southwest end of the Uintah's enginehouse. The huge wooden doors swung to the center of the track and were said to have been hung on their hinges so precisely that a slight chill wind could open them with ease. Most shop men and engine crews entered the smaller doorway between the two doors.

10

GEORGE ROSELAND COLLECTION

THE ENGINEHOUSE AT ATCHEE...

THIS WAS THE SHOP CREW ca. 1910. The Uintah Railway prided itself for repairing its own equipment, and the highly skilled men at the shop in Atchee kept everything in good working order. Here, the old two-piece doors could be swung open against large wooden doorstops hanging out from the side of the building, near the top of the doors. The sign over the door was: "NO SMOKING IN OR BEYOND THIS BUILDING." The face of the enginehouse was the farthest south portion of any building in the "yard" at Atchee, and smoking north or east of this area was off limits. Most of the railway facilities were kept as clean as possible, but coal dust, pieces of Gilsonite, and the wooden structures created an easily combustible area around the railway facilities. Unlike some of the residential buildings in Atchee, there was never a fire in Atchee that caused any casualties or loss of property. That being said, it is strange to note that three of the employees in this photograph were smoking pipes, one of them in the doorway under the sign! *What management does not know will not hurt it!*

UINTAH No. 40 and a couple of water cars were sitting on the northeast leg of the wye at Atchee. The white strip on the side of the water tank was numbered from the top down, in one-foot increments, starting a foot below the top with "1" and going to the very bottom of the tank at "16." These were measurements in feet, and a small triangular point ran up a groove between the two boards that made up the scale. This was attached to a weighted float on the inside of the tank. When the tank was full of water, the pointer would extend all the way down to the "16," showing how much water was in the tank. As the water level inside the tank fell, the float dropped, causing the pointer to rise on the outside of the tank. Notice that the water stand pipe had been moved closer to the tank and had been stabilized with rods going back to the edge of the tank. The level of the enginehouse floor was about two feet lower than the ground outside, as can be seen in this view, comparing the level of the door, compared with the windows. The large boilerhouse in the background filled the rest of the scene.

RODGER POLLEY COLLECTION

DAN PRICE COLLECTION

THE ENGINEHOUSE AT ATCHEE was 131 feet long, with two tracks, and it easily could house from four to six of the Uintah's locomotives at a time. This structure was the perfect place to take photographs of most of the engines—as they were shuffled in and out for repairs and storage. This is a great view of the tender of locomotive No. 40, with Master Mechanic Scott Shafer standing near the cab. The rivet heads that diagonally mark the tender were the result of the shop crew's effort to extend the water-carrying capacity of the trim 2-8-2—from its original slope-back design. If it were not for the seam marks, it would be difficult to tell that No. 40 was not originally built that way at the Baldwin Locomotive Works.

JIM McCOY COLLECTION

THE VINES THAT HAD been growing on the front of the machine-shop offices at Atchee had lost all their leaves, and you can see the back end of the tender of engine No. 40 was just inside of the enginehouse during the late fall of 1937. Observe the heavily sooted areas above the two doors, where smoke from the locomotives had blackened the wall and roof. The tall smokestack at the boilerhouse (behind the enginehouse) was releasing a faint trail of black smoke.

THIS ENGINEHOUSE VIEW provides you with an interesting study of the mainline curve that headed toward Baxter Pass, passing the two water tanks. Notice the section house and a nearby stock car. Engine No. 40 was sitting on the enginehouse lead track, fired-up and waiting for a train order.

ROBERT C. BOULTER COLLECTION

TAKING A BREAK, a machine-shop worker sat on a small push car stenciled "SHOP," at the southeast corner of this structure. The doors of the enginehouse had been opened, folded back on themselves for storage, and held back with a small latch on the corner trim of the building. Over the years, rain water from the eaves of the building had left the bottom two feet of the walls well weathered. A brand new eave-trough and downspout had been installed to help combat Mother Nature.

DAN PRICE COLLECTION

DAN PRICE COLLECTION

THE ENGINEHOUSE and boilerhouse in Atchee did not have insulation; warm air from the engines, boilers and equipment kept this long structure warm enough to slowly melt the snow on the roof. Huge icicles were the result, as the slightly warmed water froze again in the frigid air after running off of the eaves. The accumulation of snow had to be removed by hand from the yard tracks at the division point. The passenger platform and area around the water tanks were shoveled clean, waiting for the next snowstorm to cover them again.

RODGER POLLEY COLLECTION

WHEN THE SNOW along the railway got too heavy to be shoveled by hand, a snowplow would be pushed through the yard at Atchee to move the worst part out of the way. Snowplow K-1 was being pushed around the wye, past the south side of the machine shop, by one of the Shays. The face of the plow was covered in sheet metal. This plow had been mounted on the front of a large wooden box filled with rocks and pieces of iron for weight (called ballast).

RODGER POLLEY COLLECTION

THE SMALL WOODEN office building, off the side of the concrete machine shop, was built during the early teens. One of the large windows on the southeast side of the building was taken out and the wall was cut down from the bottom window sill to the floor, creating a doorway directly onto the shop floor. This small office was principally used by the master mechanic and the shop clerk. Pictured here during the late 1920's was Master Mechanic Scott Shafer, in the doorway, standing next to Gilbert Miller, the company clerk at Atchee. Telegraphy equipment was here, along witn all of the paper work and bookkeeping that was needed to run a machine shop and engine facility. Train orders were issued from Mack at the chief dispatcher's office, and individual orders were given from here. A woman usually was employed to help with the paper work. During the 1920's and 1930's, these clerks usually were young, single ladies from Grand Junction. Generally, they did not stay in the small mountain town for long, but while they were there, they often stirred up a lot of emotions among the shop crew, who often tripped over one another to gain the ladies' favor. A large picture window in the front of the office, a window above the door and another window on the small addition, allowed enough light in during the day, while electric lights and steam heat made the office comfortable during the short winter days.

BETTY CASSABOOM COLLECTION

THOMAS BIRD and Phyllis Boulter posed for this "boyfriend and girlfriend" photograph in the open area between the residences and the enginehouses in 1915. In the background, one of the Uintah's 0-6-2T tank engines sat in front of the enginehouse waiting to be dispatched on a passenger train. The machine shop was on the right, beyond the small master-mechanic's office. Passengers arriving in Atchee had to walk through this area on their way to the hotel.

ROBERT C. BOULTER COLLECTION

THE UINTAH'S MACHINE SHOP was built at right angles to the enginehouse. It was constructed to house the shop machinery in 1905. Like everything else that the Uintah's management did during the early years, the shop was a "real prize" for such a small railway. Outfitted with new machinery, it was capable of handling nearly any problem that could arise. Rebuilding boilers and servicing the locomotive drive gear was commonplace for the shop crew. As the years passed, more complicated jobs were tackled, increasing in complexity, including: reboilering several engines, adding superheaters, modifying coal bunkers and drive gear, building a steel-sheathed passenger car and rebuilding more than a few wrecked locomotives from the ground up, including the creation of a whole new Shay locomotive in 1933. Here, winter had set in at the enginehouse and shop building, covering everything with snow and icicles. A good clear view of the extension that connected the two buildings and the office building can be seen from this angle. The large smokestack, in the distance, actually belonged to the boilerhouse, northeast of the machine shop. Snow had covered the wye trackage in the foreground. The fence and poplar trees had been added during the late 1920's, along with a boardwalk, which extended from the residential boardwalk to the tracks in front of the office. Notice the large windowed clerestory above the shop that allowed plenty of natural light into the work area. Also, a gondola had been spotted on top of the coaling tipple, visible in the distant background, through the limbs of the trees.

BETTY CASSABOOM COLLECTION

A COUPLE HAD FUN on a child's sled during the teens, while steam and smoke were rising up from the boiler's smokestack, which was still located next to the opposite side of the machine shop. A small materials-storage shed was next to the wall of the enginehouse, and the shop offices had not yet been painted white.

MASTER MECHANIC Scott Shafer (left), General Manager Walter Rader (center) and Al Green, the ICC boiler inspector from Grand Junction, were portrayed here, standing on the front porch of Luton's store. They probably had visited the pool table for entertainment.

RODGER POLLEY COLLECTION

BILL ELLICOTT COLLECTION

THE SHOP OFFICE at Atchee was where the Uintah's engineers and train crews got their orders for the day. Except for the crews that started their day in Dragon or Mack, most of the operations of the railway started and ended in Atchee. The office was the natural place for bulletins and announcements concerning railway operations. All train crews were required to sign each announcement, verifying that they had seen what was in the bulletin. The first one of these bulletin books covered the time between September 1905 and November 1911. The following are some of the more interesting texts from that first bulletin book, now owned by Bill Ellicott:

Sept. 26, 1905

ALL CONCERNED:

Derail switching has been placed on the east end of siding at Atchee. This switch must always be set and locked for derail when not in use; and the main line switch must be set and locked for the siding when not in use. Mainline switch will serve as a derail for mainline while switching in yard.

Yours Truly,
E. A. Grove,
Superintendent

Oct. 14/1905

TO ALL CONCERNED:

Effective Oct. 16th, Railway Mail Service Route 165057 is established between Mack, Colo., and Dragon, Utah. Uintah Railway Company, Mack, Colo., will dispatch pouch daily to Dragon, Utah, by train No. 1. Dragon, Utah, will dispatch pouch daily to Mack, Colo., by train No. 2. Agents, Conductors and Motormen will see that these pouches are properly handled to destination; also any locked pouches or tie sacks for the following offices will be supplied from Dragon, Utah, effective with this change:

Dry Fork, Utah
Ft. Duchesne, Utah
Jensen, Utah
Maeser, Utah
Myton, Utah
Ouray, Utah
Randlett, Utah
Vernal, Utah
White Rocks, Utah

E. A. Grove

January 22, 1906

CIRCULAR:

Captain M. W. Cooley is hereby appointed General Manager of the company's lines. Appointment effective February first.

C. O. Baxter
Vice-president

ALL EMPLOYEES:

EMPLOYEES' RECORDS FOR FEBRUARY 1906 — A fireman has been discharged for frequenting saloons and being intoxicated while on duty. A brakeman has been suspended for 15 days on the record for frequenting saloons. A conductor has been suspended 10 days on the record for failing to bring train mail sack from Dragon Junction on train No. 2. An engineer has been suspended 10 days on the record for allowing train to slack back, thus derailing car on the South Switch [at] Sewall.

April 13, 1906

ALL EMPLOYEES:

An engineer has been suspended 15 days on the record for not seeing that the brakes of engine No. 20 was in proper shape before moving her and letting engine run into engine No. 21, and 15 days for allowing engine No. 20 to die on Baxter Pass.

E. A. Grove,
Superintendent

May 24, 1906

NOTICE —

This company is desirous of cultivating a lawn and otherwise beautifying its grounds at Mack for the benefit of its employees and the traveling public passing this way. It is hoped that all concerned will show their appreciation of this by using their best efforts to protect the same from the ravages of FOWLS and ANIMALS of ALL KINDS.

E. A. Grove,
Superintendent

THE UINTAH RAILWAY COMPANY

CAR CERTIFICATE

No. 1. ....STATION....... ....192.....

I hereby certify that I have this day personally examined ....car No...................and that the roof and sides have no loose boards, holes or cracks, or unprotected decayed spots liable to hold sparks and start a fire; that the king bolts or draft bolts are properly protected, and that there are no uncovered irons or nails projecting from the floor or sides of the car which might injure packages of explosives; also, that the floor is in good condition and has this day been cleanly swept before the car was loaded; that I have examined all the axle boxes and that they are properly covered, packed and oiled, and that the air-brakes and hand-brakes are in condition for service.

....
Railway Employe Inspecting Car

No. 2. ....STATION........ .... 192....

I hereby certify that I have this day personally examined the above car, that the floor is in good condition and has been cleanly swept, and that the roof and sides have no loose boards, holes, cracks or unprotected decayed spots liable to hold sparks and start a fire; that the king bolts and draft bolts are protected, and that there are no uncovered irons or nails projecting from the floor or sides of the car which might injure packages of explosives; that the explosives in this car have been loaded and stayed, and that the car has been placarded according to paragraphs 1661, 1666, and 1674 to 1683 inclusive, of the Regulations for the Transportation of Explosives prescribed by the Interstate Commerce Commission; that the doors fit or have been stripped so that sparks cannot get in at the joints or bottom.

....
Shipper or Duly Authorized Agent

....
Railway Employe Inspecting Loading and Staying

NOTE.—Both certificates must be signed. Certificate No. 1 by the representative of the carrier. For all shipments loaded by the shipper, he or his authorized agent, must sign certificate No. 2, and the representative of the carrier must certify as to loading or staying and general condition. When the car is not loaded by the shipper, certificate No. 2 must be signed only by the representative of the carrier. A shipper should decline to use a car not in proper condition.

RODGER POLLEY COLLECTION

YOUNG SCOTT SHAFER was leaning back in a chair in the shadow of the section house at Atchee. The Uintah always had that small railway feel, because trains did not run on Sundays and almost never at night. Although the living conditions were somewhat basic at times, the fact that an employee could count on being home nearly every night and on Sundays was a big incentive for many of the men to stay on with the Uintah, rather than look for greener pastures elsewhere. In the background was the small shed where handcars were stationed.

April 9, 1907

TO ALL CONCERNED:

It has been customary for section foreman and their men to use the telephone placed in the section houses, to carry on social conversations with their friends along the line, and it is carried to such an extent that it has become a nuisance, the line always being in use when it is required for more important business. In the future, it will not be permissible for the conversations to be carried on unless they are paid or responsible for the collection of tolls covering any such conversations.

Yours Truly,
E. A. Grove,
Superintendent

July 23, 1907

ALL CONCERNED:

Where flatcars are unloaded which have stakes in the pockets, please see that the stakes are left on the car and returned to Mack.

Yours Truly,
E. A. Grove,
Superintendent

August 7, 1907

ALL CONCERNED:

A conductor has been suspended 15 days on the record for flattening wheels on a caboose, July 31, 1907.

E. A. Grove,
Superintendent

Sept. 8, 1907

All Engineers and Conductors and All concerned:

You will not allow anyone, other than employees, to ride on engines without permission from this office. Passengers will not be allowed on freight trains between McAndrews and Atchee without permission from this office.

E. A. Grove,
Superintendent

Oct. 22, 1907

ALL CONCERNED:

In the future, do not move cars loaded with loose Gilsonite from Dragon mine unless the same are covered with canvas. Reporting the same to this office.

Yours Truly,
E. A. Grove,
Superintendent

July 17, 1909

ALL PASSENGER CONDUCTORS
ALL FREIGHT CONDUCTORS

Effective at once, passenger conductors will report to this office (Mack by means of telephone) at Atchee, Baxter Pass and Wendella each day without fail, and freight conductors will report at Salt Wash, Carbonera, Atchee, Shale, Baxter Pass, Columbine and Wendella unless otherwise instructed, giving arriving time of train and probable duration of stop, and when detained over five minutes report to dispatcher again before departing.

D. E. Falvey
Superintendent

DAN PRICE COLLECTION

A QUIET DAY AT ATCHEE found articulated locomotive No. 50 sitting quietly inside the west stall of the enginehouse. One of the Uintah Railway's steel water cars was sitting on the wye behind the water tank. The yard was virtually spotless and weedless in every direction you looked. The machine shop and office were to the right. The office had a small section appended to the building's side. During the 1920's, a small addition to the enginehouse was built in the same area, and the steel tank, with the pipe running from it, were added.

Mar. 24, 1910

TO ALL CONCERNED:

Engine men must be particular to see that ASH PANS on their engines are kept closed, so that bridges will not be liable to catch fire.

D. E. Falvey
Superintendent

TO ALL CONCERNED:
EFFECTIVE MONDAY, APRIL 24, 1911 —

Mr. W. C. Watson has been appointed Engineer of Construction. All employees in the Engineer's Department will report to him for instruction.

Captain Cooley
General Manager

Form 2110-10-27-1000.

THE UINTAH RAILWAY COMPANY

BRAKE AND SIGNAL EQUIPMENT

Engine No. 40

Tested and Cleaned at Atchee, Colorado Shop

Air Compressor, Orifice Test	Date 11-18-38 192
Main Reservoir, Hydrostatic Test	Date 8-18-38 192
Air Gauges Tested	Date 11-18-38 192
Distributing Valve Cleaned	Date not used 192
Control Valves Cleaned	Date 8-18-38 192
Triple Valves Cleaned	Date 8-18-38 192
Reducing Valves Cleaned	Date 8-18-38 192
Straight Air Double Check Valves Cleaned	Date not used 192
Dirt Collectors Cleaned	Date not used 192
Brake Cylinders Cleaned	Date 8-18-38 192

RODGER POLLEY COLLECTION

RODGER POLLEY COLLECTION

THE SHOP CREW was photographed in the mid-1920's. This was when the Uintah Railway was in its heyday, and there was plenty of work to be done. The fleet of engines and cars was getting old and repairs were needed on a regular basis. The shop crew as pictured here was as big as it ever got, and it dwindled during the Great Depression. As people left over time, they were not replaced. The ladies in the photograph helped with things in the office, including the telegraph. Most of them were hired for only a short while, before moving back to a larger town. Ruby Luton could always be counted on to help with the duties and was a temporary fill-in at the shops year after year. In her words: "[The ladies] didn't know what to think of such a crude place and didn't stay around long."

RICHARD A. RONZIO COLLECTION

THIS VIEW OF ATCHEE presents a different angle, looking off the coal-chute ramp toward the concrete machine shop. The machine-shop walls were made of thick concrete that had been poured in layers all the way to the roof Notice the wye turnouts at this location.

ROBERT C. BOULTER COLLECTION

THE COMPANY CLERK at the Atchee shop, Glen Griffith, was standing in front of a pile of odds and ends from the machine shop. This photograph was taken from just outside the doors of the shop, looking northeast toward the coal-tipple trestle in the background. To the left was the boiler of the Uintah second Engine 11, resting on the wye tracks that ran behind the machine shop.

W. C. LARSON COLLECTION

A PLUME OF SMOKE was rising from one of the Consolidation engines, waiting to haul a water train south to Mack. The Uintah's three-tank water cars carried more water than the steel tank cars that replaced them, but they required much more maintenance than the steel cars. They had to be rebuilt and sealed on a regular basis, as sometimes they were found to be leaking out more than a quarter of their water before reaching Mack. The residential portion of Atchee had spread out along the valley, and at the mouth of the canyon—just in front of the old school building—you can make out one of the older passenger cars, now being used as a residence.

PORTER PHILLIPS COLLECTION

UINTAH CABOOSE No. 02 had been spotted on the back side of the wye, behind the enginehouse. This small four-wheel caboose was one of two that were used by the Uintah and rebuilt on several occasions. They both had cupolas added later. This photograph probably was taken during the late teens. The lean-to on the side measured 18 feet long by 8 feet wide. It was used as a storage rack for stock pieces of metal bars, rods and pipe for use in the enginehouse and shops. This later was moved, during the 1920's. The large white building in the far background was on the Kelley's ranch.

RODGER POLLEY COLLECTION

THE PROPOSED EXTENSION of the Uintah Railway to the mines at Bonanza, Utah, was the catalyst for many surveys in that area and on Baxter Pass. Tunneling the pass, standard-gauging the line and extending the rails to the Gilsonite mines north of the White River were all long-term goals for the Uintah's management and the parent mining company. The surveyors would be kept busy for a couple of years around 1922, doing many different jobs on the grade over Baxter Pass and the extension north from Watson. The main means of transportation was a railcar built at the Atchee shops, using the body and engine from a 1916 Buick touring car. Needing the ability to quickly get from point to point on the railway, and with the old Mack Brothers No. 50 railcar nearing the end of its useful life, the Buick was built using parts from No. 50, including the front truck and other pieces. This picture of the Buick was taken in front of the enginehouse on the tracks going to the stalls. The man in the picture was Bruce Angus, who had been hired on to the Uintah as part of the survey crew to get the surveyors from Atchee to their destination on the pass via the rail car. Mr. Angus was quite the eligible bachelor at the time, and in this picture he is seen (from left to right) with: Katherine Rogers, Elizabeth Dow (later, Mrs. Elizabeth Angus) and Elizabeth Smith. They were getting ready to go on a Sunday "excursion." This particular trip took them over Baxter Pass to the Gilsonite mines at Rainbow. There, they met with friends at Rainbow and took a tour of one of the mines on the Colorow Branch. It was mid-March of 1923, and everyone was dressed for the cold trip over the top of the still-frozen mountain.

PORTER PHILLIPS COLLECTION

THIS ENLARGEMENT of the caboose reveals more details. The lettering on the side of the caboose reads: "UINTAH" on the left side of the window, "RY" on the right side and "CABOOSE 02" underneath the window.

RODGER POLLEY COLLECTION

THE CAMERA LENS WAS AIMED toward the engine facilities at Atchee for this photograph—to the east—which was taken during the 1930's. The length of the big enginehouse stands out beyond the two water tanks. Nearly hidden on the wye stood the American Hoist & Ditcher crane, emitting smoke from its smokestack. A tool shed had been placed beyond Bridge 28-A, in the foreground. A section house had been situated on the embankment, above the wash. Both the older schoolhouse and new one appear at opposite ends of Atchee. The turnout in the foreground and the mainline showed evidence of locomotive firebox ashes being dropped on the ties. A spur was constructed at the same time the boilerhouse was built, in order to store snow-fighting equipment. Cattle guards had been constructed to protect the town from wandering livestock.

RODGER POLLEY COLLECTION

A LIGHT STREAK on this picture hinders the detail on the American Hoist & Ditcher steam shovel, spotted next to the boilerhouse in Atchee not long after its arrival in 1924. The sliding white-canvas curtain was used to regulate the temperature in the "cab" of the unit and keep dust from the shovel from blowing in on the machinery. When not in use, the ditcher was kept on the storage track east of the coal chute; however, it later was kept on a short spur on the other side of Bridge 28-A.

RODGER POLLEY COLLECTION

THE UINTAH RAILWAY'S American Hoist & Ditcher steam shovel was bought specifically to combat the slide near Shale Tank south of Baxter Pass. Here, we see it heading out of town under its own power, going toward a work site on the pass. The bucket had been lowered into a gondola for safety while traveling. Notice the connecting rods on the sides of the truck, which enabled all of the wheels to be used for traction. The engineer had to keep a close eye on the conditions of the track in front of the slow-moving ditcher to prevent a derailment. The top-heavy machine could easily get off-balance, and great care was taken to keep it from leaving the rails. This was one more reason for the use of the gondola. With the gondola, the weight was distributed on the gondola's trucks.

ROY F. BLACKBURN PHOTO – JIM McCOY COLLECTION

WATER FOR ATCHEE was supplied by its own pipeline, which ran to a spring located a few miles away, at a natural aquifer in the shale in the mountains northeast of town. A small iron pipeline was run from the spring down West Salt Wash to Atchee to fill the water tanks. The first water tank at the division point was designed to meet the needs of the railway and town. The tank measured 18 feet in diameter and stood 16 feet tall, on a 14-foot-tall timber frame. This water tank was the most distinctive feature of the "skyline" at this location. The first tank was insufficient to hold enough water for the growing town and railway. By 1919, the Uintah had built another tank next to the original. This one was 24 feet in diameter and stood 16 feet tall, built on a concrete base. It doubled the water capacity for the town and enabled the railway and shops to run more efficiently. It also allowed the residents to "waste" a little water on lawns and trees. This picture was taken in 1939.

RODGER POLLEY COLLECTION

AT SOME POINT during the late 1920's, Atchee's secondary water tank burst, littering the ground with splintered wood—washing the entire area clean. This picture was taken from near the mainline, looking at the older water tank and the remains of the latest one, piled on the ground and across the track. Although this seems to be a small tragedy in the scheme of things today, the collapse had major implications. Not only was there lumber to be cleared from the track, but the tank supplied water to the stationary boiler in the boilerhouse. Although the other water tank could be utilized to provide water to the boiler, as well as enough water to supply the needs of the engines, the towns of Mack and Carbonera had to be planned for. It would take several weeks to rebuild the water tank, and the three communities were quickly informed in no uncertain terms that water was not to be wasted. To the left was the section house. One of the first buildings built in Atchee, this section house was utilized as a bunkhouse for the railway's employees responsible for track maintenance from Carbonera to Baxter Pass. In the background, from left to right, was the tool shed, pipe shed and boilerhouse.

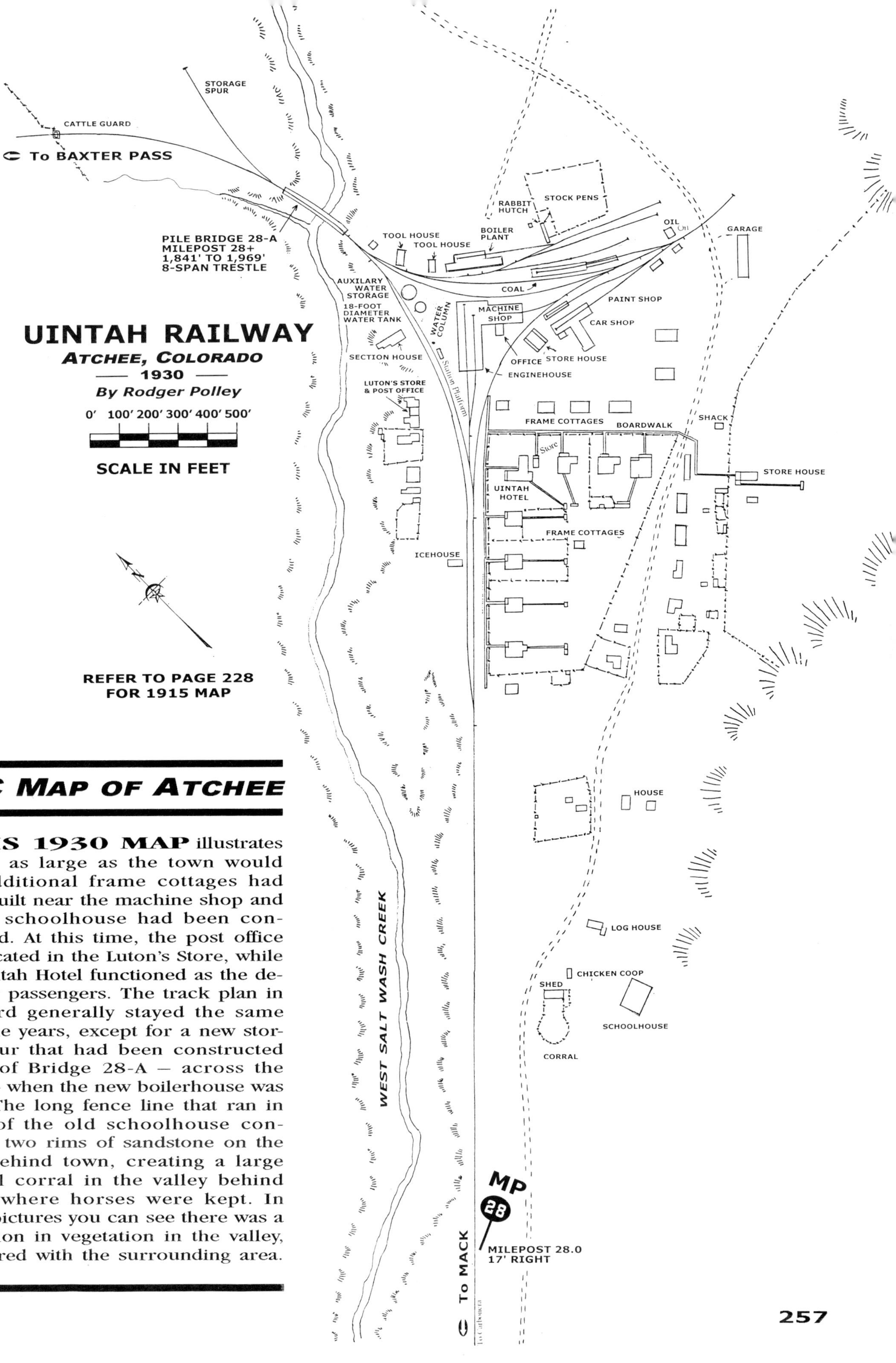

ICC MAP OF ATCHEE

THIS 1930 MAP illustrates Atchee as large as the town would get. Additional frame cottages had been built near the machine shop and a new schoolhouse had been constructed. At this time, the post office was located in the Luton's Store, while the Uintah Hotel functioned as the depot for passengers. The track plan in the yard generally stayed the same over the years, except for a new storage spur that had been constructed north of Bridge 28-A – across the wash – when the new boilerhouse was built. The long fence line that ran in front of the old schoolhouse connected two rims of sandstone on the hills behind town, creating a large natural corral in the valley behind town, where horses were kept. In some pictures you can see there was a reduction in vegetation in the valley, compared with the surrounding area.

RICHARD A. RONZIO COLLECTION

THE TRACK IN FRONT of the car shop often was used for rebuilding and repairs on cars when the interior of the car shop was filled with other bad-order cars. This passenger car had been worked on, possibly having the running gear repaired, as noted by the wooden supports under the sides of the car. The rails that were perpendicular to the siding went to a storage area for wheelsets.

RICHARD A. RONZIO PHOTO

AS HE CLIMBED THE SLOPE east of the site of Atchee during June of 1959, Richard A. Ronzio photographed this scene bathed in morning sunshine. Ruins of the machine shop and the row of trees where the Uintah Hotel and cottages once stood are all that remained of the division point.

52

ROBERT C. BOULTER COLLECTION

THIS PANORAMA OF ATCHEE was taken from the northwest side of town, looking at the railway yard and town, spread out to the right during the mid-1930's. At the far right was the white school building, sitting at the base of the mountain. In the foreground were the remains of passenger car No. 51, being used as a residence, with most of the windows boarded up, a door on this side and the back platform turned into a bathroom. The myriad of buildings with connecting roofs were all part of the Lutons' store, U.S. Post Office and home. Packaging from the store and spoiled goods were tossed over the embankment behind the store. On one occasion, a group of children found and ate a whole box of discarded "Ex-Lax" laxative—believing it to be chocolate! There were many upset mothers in Atchee that night! The poplar trees throughout the town owed their presence to Ruby Luton. Although many trees had been started in the past, there was little water for them, and they usually died within a year. This changed during the late 1920's, when Ruby took it upon herself to get everyone in town stirred up about the trees. The whole town was behind her in this cause. Ruby wrote letters to General Manager Sprague in Mack and even to company officials back East. As the old saying goes, "the squeaky gate gets the grease," and her tireless campaign eventually paid off. General Manager Sprague showed up on a train with a flatcar loaded with young saplings, and he said she was responsible for getting the trees planted without any help from railway employees. After management headed back to Mack, everyone in town turned out to help plant the young trees and get them watered. Being company property, water was allowed to be used to irrigate the trees and keep them healthy.

OTTO C. PERRY PHOTO – DENVER PUBLIC LIBRARY, WESTERN HISTORY DEPARTMENT

ALTHOUGH RAILCAR 52 made trips over the "hill" to Dragon, this vehicle spent most of its time working as a shuttle between Mack and Atchee. This is how the railcar looked on the wye in front of the enginehouse and section house at Atchee on April 9, 1939. The starter crank was hanging below the radiator. Sometimes it was hard to start a vehicle like this on cold winter days, unless it was kept in a warm place overnight, like the enginehouse.

ROBERT C. BOULTER COLLECTION

AN ENLARGEMENT of the previous image shows the railway yard during the mid-1930's. The roof of the long enginehouse had been repaired near the center, with a large patch on the roof. A light breeze from down the canyon was blowing the smoke from the section house and boiler stack northeast, away from town. The air in the small valley where Atchee sat could get stale during the winter, and smoke from the boiler and all of the engines moving around got thick on many occasions. Along the base of the hillside in the background was the row of small dugout garages, which were built by the railway men to house their automobiles. Considering themselves to be lucky to have jobs during the Great Depression, many of the Uintah's employees were content. While many railroaders around the country were being laid off or were wrapped up in some labor dispute, Uintah employees were financially able to buy new automobiles. The overhead pipe from the boilerhouse to the shops was put in sometime in the 1920's, in favor of an older, smaller pipe that was underground and prone to rupture under the yard tracks. The area where the boilerhouse, tool shed and storage shed were built originally was the place where a short spur had been laid, which was used to store the snowplows and snow spreader. When the boilerhouse was built, this spur was removed and rebuilt on the other side of Bridge 28-A, across West Salt Wash. Notice the sliding door on the tool house and the gondola full of coal, spotted at the coal tipple, on the back side of the boilerhouse.

RICHARD A. RONZIO COLLECTION

THE PHOTOGRAPHER was looking toward the southwest for this picture of the storage yard in Atchee. The community automobile garage can be seen, as a long row of sheds. In the background was the old schoolhouse, being used as a storage building before it burned down. It is unclear as to where the large boiler came from, but it was possibly one of two boilers that the shop at Atchee had worn out over the years. The long caboose-like boxcar was most likely car No. T-12, used specifically for wrecks. It contained all of the necessary tools and equipment employed for salvaging wrecked locomotives and freight cars.

BAXTER PASS, showing many scars on the mountain from the right-of-way of the Uintah Railway, can be observed from this vantage point—as the steep grade snaked its way up the hillside. Notice near the center of the photograph the slide at Moro Castle curve, where the constant battle against nature has left tons of dirt pushed over the side of the mountain. Several gondolas and other maintenance-of-way equipment were sitting on the spur at the right of Bridge 28-A. Rabbits were the reason why the buildings in the foreground (at the right) were built. Gilbert Miller, the company clerk at the Atchee machine shop for many years, decided to make some extra money breeding rabbits. He constructed a huge pen and countless rabbit hutches for breeding and keeping the rabbits. He mail-ordered a bunch of young bunnies, brought in on the train. The rabbits did well, and did as all rabbits do, creating a vast "herd" of hundreds of rabbits within a few years. As with any "get-rich-quick" scheme, especially in the heart of the Great Depression, the real intent of the advertisements for bunnies flopped as the rabbits matured. The ads were designed to sell bunnies in great numbers, but there was no one who wanted full-grown rabbits! The rabbits finally were sold at a great loss. Mr. Miller was out all the time and money he had put into the bunny venture—*sad but wiser!*

ROBERT C. BOULTER COLLECTION

RICHARD A. RONZIO COLLECTION

THIS PHOTOGRAPH was taken from the same location, looking toward the machine shop and storage tracks from the pasture area east of town during the late 1920's. Flocks of sheep were brought up the canyons of the Book Cliffs during the spring and summer to graze on the natural grass in the valleys. A man stood with a small lamb in front of the Uintah's complex consisting of shop and engine facilities. Two gondolas full of coal had been left on the coal-tipple track for both the boilerhouse and the main coal tipple. The boilerhouse tipple was hidden behind a covered corral, in the foreground, which was used to breed and raise rabbits by the company clerk.

RICHARD A. RONZIO COLLECTION

OTHER THAN GILSONITE ORE, water accounted for the largest number of carloads heading south to Mack from Atchee. Here, Engine 40 was sitting next to the water tank as two of the steel water cars were filled from the main spout. Behind the main water tank, you can see the machine shop's large water tank, which rested on the ground—including its inspection ladder.

RODGER POLLEY COLLECTION

ANOTHER VIEW OF ENGINE No. 40 portrays this handsome Mikado, while Uintah water cars were being filled on September 11, 1938.

RODGER POLLEY COLLECTION

A FIREMAN STOOD on the back of Engine 40's tender, holding down the water spout as it filled the tank. This trim Mikado and the two water cars were both in good shape, with paint jobs that match, complete with pin-striping on the locomotive and water cars. Notice the small three-inch pipe in front of the engine. This was known as a "water stand" and was a direct line to the water tower. The stand pipe had a pivot at the base that could be swung out over the rails. These were built primarily for filling water in the little passenger side-tank locomotives, Nos. 20 and 21. Near the ground, you can see a removable "T" handle that was attached to a valve underground, which opened the water flow to the water column. In the background was the doorway to the boilerhouse and a large smokestack for the boiler.

DAN ABBOTT COLLECTION

UINTAH ENGINE No. 40, in perfect shape, was photographed here as she sat on the wye, next to the enginehouse on September 11, 1938.

OTTO C. PERRY PHOTO
DENVER PUBLIC LIBRARY, WESTERN HISTORY DEPARTMENT

HERE IS ENGINE 12 before she had been re-boilered at the Atchee machine shop in 1924. She was photographed ca. 1920, with the older boiler. The only real difference between the original boiler and the newer one was that the smokebox was shortened by a couple of feet from its original length. The tender also was reworked at this time, adding a taller and longer skirt around the top to hold more coal. Look at the photograph of this little Consolidation printed on the next page to see the difference in the tender skirt. The machine shop and enginehouse are visible in the background of the scene on this page. Observe the overhead steam pipe, which traversed the yard to the boilerhouse. Electric wires then ran back across to the boilerhouse, along the bottom of the steam pipe, from the dynamo in the machine shop.

RICHARD A. RONZIO COLLECTION

THE THREE-WAY stub switch at the back of the wye had been installed when the trestle was built, to reach the top of the coal bins. The actual wye could be used by entering it by way of one the two switches below the three-way switch, which was used to reach either to the coal bins or the car shop. Caboose No. 2 and a maintenance-of-way car (with a cupola) were spotted on a storage spur.

A TRAIN CREW had posed for a picture beside the engineer's side of rarely photographed Shay No. 2—sitting on the mainline at the south end of town. For good luck, the train crew had hung a horseshoe on the post of the grab-iron railing near the front of the smokebox. Superstition and tradition both played parts on how train crews often decorated their locomotives, although, as some might point out, the luck was likely to run out with the shoe in this position. This former Argentine Central locomotive came from the high-mountain country of Colorado; however, she was scrapped in April of 1928.

RICHARD A. RONZIO COLLECTION

RICHARD H. KINDIG COLLECTION

THE UINTAH'S No. 12 sat next to the machine shop ca. 1933, after being serviced. Originally, she was a Florence & Cripple Creek Railroad engine, No. 10 (which later became the Cripple Creek & Colorado Springs Railroad's No. 36) This 2-8-0 was sold to the Uintah during August of 1917, and during 1924, a new boiler was applied to the engine at the shop in Atchee. On the northeast side of the machine shop, a parallel track formed the opposite side of the wye, and was used for coaling, next to the coal tipple. This location was used as an ash pit, for dropping cinders and ash out of the fireboxes of the engines. During the winter, these ashes were hauled to West Salt Wash by wheelbarrow and dumped over the side. During most of the year; however, they were left in piles between the two tracks, waiting to be loaded onto gondolas to be taken along the right-of-way and used as ballast. This was a labor-intensive job, but nothing went to waste on the Uintah, especially something as effective as cinders and ash to keep the roadbed from eroding. In the background was the Uintah's car shop. Built around 1910 for the expansion of the Uintah's fleet of flatcars, this large building was used to maintain and construct freight cars. Utilizing a complete woodworking shop, the machine-shop crew was split into two divisions, one for engines and one for freight cars and maintenance-of-way equipment.

JOHN AND HAZEL CARSON COLLECTION

ORIGINALLY PURCHASED in April of 1905, Shay No. 3 was used for the first five years of the railway's operations to move trains from Atchee to Dragon, Utah, and back. Shays 1, 2 and 3 were operated exclusively on that part of the line except for the regular passenger run with the small tank engines. Nos. 2 and 3 were much larger than the original No. 1 and saw the majority of the work, running extra to the top of the hill. No. 1 was used as the engine normally used on most maintenance-of-way trains along the entire line. No. 3 was standing in Atchee, showing an excellent view of the Shay's complicated driving gear. The steam was delivered to the three cylinders on the side of the engine, which in turn, drove three rods that were attached to a crankshaft at the bottom of the "engine." Power was then distributed down the sides of the engine via the long drive shafts to both of the wheelsets. These drive shafts had large flexible joints on them at the engine's end and at the wheels, along with a slip joint, allowing the Shay to maneuver around very sharp curves, while not causing the drive gear to bind. At the wheels, a large beveled gear matched up with gears on the side of the wheels, providing an all-wheel-drive system—very effective on steep grades and on trackage that a rod engine might have trouble getting over.

UINTAH R.R. was the lettering applied to the side of Uintah Shay No. 4 by the Argentine Central Railway before her arrival on the Uintah Railway. The engine was shipped from Silver Plume, Colorado, during 1909. Here, she sat in the yard at Atchee. The gentleman in front of the cab (with the watch chain hanging down) was Alex G. Beverly, superintendent of the Argentine Central, and this Shay had been their No. 2. He accompanied the locomotive to her new home and was familiarizing the Uintah employees with the engine's maintenance requirements. The incorrect "R.R." lettering scheme stayed on the engine for a while, probably changed at the time new paint was applied after the cab was rebuilt.

DAN ABBOTT COLLECTION

ROY ENO, JR. COLLECTION

IN THIS SCENE, SHAY No. 3 was being used as the helper engine on a train full of empty flatcars and a load of ties destined for somewhere on the other side of Baxter Pass. Coupled to the rear end of the train, the consist would be at the top of the 7.5-percent grade in no time. Engineer Roy Eno was in the cab, as it looked with the open cupola on top—after it was rebuilt following a wreck on Baxter Pass in 1909. Other changes had taken place over the years, as the extra pieces of metal holding the coal back on the tender and the metal shroud over the top of the cylinders are missing. Notice the small steam exhaust behind the smokestack.

RODGER POLLEY COLLECTION

COALED AND READY for the assault on Baxter Pass, the Uintah's Shay No. 5 was resting beside the coal chute in Atchee. This was the fireman's side, showing the location of the air-pump above the running board. An air-reservoir tank was positioned beneath it.

ROBERT C. BOULTER COLLECTION

SHAY GEARED ENGINE No. 5 sat next to the machine shop, with caboose No. 2 behind her, not long after No. 5 was purchased. This was before the engine got a new boiler in 1930. In this picture, it can be seen that some modifications had been done to the exterior of the locomotive, including a new steel sheathing over the air reservoir, to allow easier access by the crew when walking next to the boiler.

A CLOSE-UP PICTURE was taken of the drive gear and cylinders of Shay No. 5, with its early boiler. In this view, you can make out the crankshaft and how it was split into thirds. The "U-shaped" section of the crankshaft under the cylinder to the right is at its farthest stroke pointing directly down toward the ground. The crankshaft in the middle was pointed toward the camera, while the section on the left was away from the camera, almost obscured in the shadow.

ROBERT C. BOULTER COLLECTION

RICHARD H. KINDIG COLLECTION

SHAY ENGINE No. 5 was in fine shape in 1937. The machine shop is in the background of this view. This geared engine had been totally rebuilt in 1930, with an entirely new boiler, becoming a proving ground for the building of Shay No. 7 from the ground up in 1933. The new boiler was larger than the old one and was designed around all the technology of the time, replacing the old 1906 boiler that was on it when it was delivered to Colorado's Argentine Central Railway as their No. 3, and later sold to the Uintah during 1910. Sitting in front of this Shay is the tender of Engine 12. The shop's main building was sheathed in tin siding. A large many-paned window can be seen through the space between the two locomotives, as well as the smokestack on the roof.

PHOTO AND LIMA FORM FROM CHRIS H. UEBERROTH COLLECTION

LIMA LOCOMOTIVE WORKS SHAY geared engine No. 6 of the Uintah Railway. This was a 50-ton two-truck Shay built in 1920, and she was rated with a 21,885-pound tractive effort.

LIMA LOCOMOTIVE WORKS, INCORPORATED
LIMA, OHIO

Class: 50-Ton 2-Truck Shay Geared

Road No. 6

Built for UINTAH RAILWAY CO.

GAUGE OF TRACK	DRIVING WHEEL DIAMETER	FUEL KIND	CYLINDERS			BOILER		FIREBOX	
			NO.	DIAMETER	STROKE	DIAMETER	PRESSURE	LENGTH	WIDTH
36"	33"	SOFT COAL	3	11"	12"	44½"	200 LBS.	72½"	46¼"

WHEEL BASE		MAXIMUM TRACTIVE POWER	FACTOR OF ADHESION	TUBES		
DRIVING	ENGINE			NUMBER	DIAMETER	LENGTH
4'-4"	30'-2"	21885 LBS.	4.93	151	2"	8'-11"

AVERAGE WEIGHT IN WORKING ORDER, POUNDS		GRATE AREA SQ. FT.	HEATING SURFACES, SQUARE FEET		
ON DRIVERS	TOTAL ENGINE		TUBES	FIREBOX	TOTAL
108000	108000	23.3	702	96	798

Capacity, Water 2500 Gallons

Fuel 4 Tons

Negative No. 231

RODGER POLLEY COLLECTION

THE FIREMAN'S SIDE of Shay No. 6 was photographed on the wye at Atchee. She was the second-largest Shay on the roster, next to the No. 7, and also the second-shortest-lived engine the Uintah owned. She was delivered new from Lima Locomotive Works in February of 1920, and was the backbone of the motive power, doing the hauling over Baxter Pass for the next seven years. When the articulateds arrived, No. 6 was one of the first engines to be replaced, and she was still relatively young, demanding a better price than some of the older engines in the fleet at the time.

ROBERT C. BOULTER COLLECTION

THE UINTAH RAILWAY had a large inventory of new parts for its fleet of Shay locomotives, and with the scrapping of No. 6 in 1927, Nos. 1 and 4 in 1928, and the most recent scrapping of No. 3, in September 1933, there were plenty of spare used parts sitting around the shops and railway yard. A new boiler, No. R14298, was ordered from Lima in 1933 and shipped to Mack, where it was unloaded and brought to Atchee. Utilizing cast pieces from the cylinders and portions of the frame from No. 3, the crew constructed a new engine from the ground up. New large 33-inch wheels were ordered, with all of the necessary bearings and moving parts to make the drive system brand new. In this view, taken during December of 1933, the Atchee crew was standing in front of their newly built Shay engine. From left to right are: Doc Dunlap, Bill Carr, Bob Wilson, Bill Mort, Roy McCoy (in the cab), Floyd Kendall, Bud Karn, Glen Griffen, Joe Zellers, Robert Hester, Bob Boulter's grandfather, Ted Osborn, Harry Dajorning, Bruce Angus, Morris and little Johnny High, Tom Bird, Jim Gold, General Manager Walter Rader and Master Mechanic Scott Shafer.

DR. LESTER L. WILLIAMS COLLECTION

RICHARD A. RONZIO COLLECTION

WEIGHING 59 TONS, No. 7 was the largest Uintah Shay by five tons and much larger than the relatively small Shay No. 1, at 37 tons. Running boards under the front pilot, a headlight viser and a cinder catcher on the smokestack was added before this photograph of the Shay was taken in November of 1938, with a full load of coal, ready for a day's work. Nos. 7 and 5 were the only Shays remaining on the line during the last years, working most of the time as switchers in Atchee and running extras on Baxter Pass.

SHAY ENGINE No. 7 had been spotted next to the coal chutes on September 3, 1934, showing that a different kind of cinder catcher had been installed atop the smokestack. The level of the original coal bins can plainly be seen in the background, about the height of the cab and sand dome of the locomotive. The newer addition contained chutes with hinged aprons, to reach locomotive tenders. In the background you can see a glimpse of the Uintah's snow-spreader attached to its familiar gondola.

DENVER PUBLIC LIBRARY COLLECTION

ROY ENO, JR. COLLECTION

LOCOMOTIVE No. 7 was constructed in the Atchee shops. She was used as a switcher and ran extra to Baxter Pass, and this Shay would never venture farther than the top of the pass on the Atchee side. Seen here in her normal role as a switcher, this engine paused for a picture on the track just south of the enginehouse.

ANOTHER PORTRAYAL of the fireman's side of Shay geared engine No. 7 was produced during November of 1938. "Drain Daily" had been hand painted on the side of the air-reservoir tank to remind the crew to prevent ice from forming in the airhoses.

ROY F. BLACKBURN PHOTO – RICHARD A. RONZIO COLLECTION

OTTO C. PERRY PHOTO
DENVER PUBLIC LIBRARY, WESTERN HISTORY DEPARTMENT

THE FIREMAN'S SIDE of engine No. 7 was photographed sitting next to the end of the boiler house on May 9, 1939. No wood on the cab and very little on the end sills made this a very "modern-looking" Shay for the time period. The overhead steam piping that ran from the boilerhouse to the shop was supported on this end by a large mast that had a cable running over the top of it for support, much like a suspension bridge. The steam pipe, the guide wires for the tall smokestack, as well as various electric and telephone lines made this part of the yard a crisscross of activity.

ON A COLD DAY in November of 1938, Shay geared engine No. 7 sat simmering on the track that went to the enginehouse. This engine was the largest geared engine on the roster and included all the improvements that Lima and the shop crew at Atchee could design, with an entirely enclosed cab and enlarged tender, which were hand-built at Atchee.

JOHN W. MAXWELL, JR. COLLECTION

RICHARD RONZIO COLLECTION

THE YEAR WAS 1926, and Shay No. 5 and tank-engine No. 21 were coupled together, ready to pull a special two-passenger-car train over Baxter Pass. The lead locomotive was the helper and would be cut off the train at the summit of the pass to return to Atchee. This photograph was taken as the camera was pointed west, with the cupola of one of the Uintah's cabooses and the top of the section house in view on the other side of Shay No. 5.

Form 2130.

THE UINTAH RAILWAY COMPANY

FOREIGN LINE TICKETS HONORED

THE D. F. HOECKEL B. B. & L. CO., DENVER 12961

DATE		ISSUING ROAD	FROM	TO	TICKET		UINTAH RY. PROP'N	REPORTED
Month	Year				Form	No.		

RODGER POLLEY COLLECTION

FREDA AND FRANCIS RIZZARI COLLECTION
COURTESY OF ED AND NANCY BATHKE

THIS ENGINEHOUSE VIEW shows how the east side doors had been remodeled and widened, with a narrow middle section hinged on the right side door, which could be opened by itself. As the age of sleek steel-sided passenger equipment was spreading across the mainlines of America, it was only fitting for the Uintah to produce another state-of-the-art piece of narrow-gauge equipment. Utilizing the body of one of the old ex-D&RG Pullman cars, private car No. 50, named the "Columbine," became the new Uintah combination car, shown here shortly after completion in 1934. Utilizing the frame and running gear of the original car, the combine was totally remodeled inside and out. The interior was redesigned and a new interior was replaced through out. Steel sheathing covered the entire exterior of the car. The old clerestory-style roof was replaced and modern windows were installed, seven on the left side and six on the right side. This would be the Uintah's only real passenger equipment for the next five years. Special occasions would bring No. 25 out for service and business car No. B-8 was only used by management and special guests. Master Mechanic Scott Shafer stood at the back platform, justifiably proud of another achievement from his shop. While many small carriers around the country were cutting back or dropping passenger service all together, the Uintah Railway's passenger service remained intact to serve the towns the railway itself had created.

ED AND NANCY BATHKE COLLECTION

THE REGULARLY SCHEDULED MIXED TRAIN had stopped at the platform on April 12, 1937, to deliver supplies to Atchee. Just about everything needed in town was brought in "mail order" from Mack, Fruita and Grand Junction. Included in this batch of parcels is a new drive shaft for one of the Shays. combine No. 50 was about the only piece of equipment to provide passenger service through the 1930's, and its arrival was anticipated by the townspeople along the line, as it went about its job delivering small shipments of freight (L-C-L) and an occasional passenger. Spring came late to the valley around Atchee as the cold temperatures from the pass and the small amounts of sunlight would keep the area considerably cooler than Mack, 28 miles away.

A SUMMER OF NEGLECT had allowed the weeds to mature around the yard and boilerhouse by November 1939. In the extreme distance, smoke from the scrap train was lingering across Baxter Pass as rail was being pulled up near the top of the pass by this time. Before the end of the month, everything in this picture would be gone. Boxcar No. 213 was still in prime shape, but gondola No. 309, which had been used by the scrappers to haul iron, showed that the wood sides had been badly damaged.

CHRIS H. UEBERROTH COLLECTION

U.RY
200

U.RY
213
U.RY
213
213

CHRIS H. UEBERROTH COLLECTION

BY NOVEMBER OF 1938, gondolas No. 306 and No. 302 were starting to show their age, with 34 years of constant use, moving thousands of tons of coal and uncountable hours of maintenance recorded. Sitting in front of the enginehouse on a cold winter day, gondola No. 306 was in need of new paint. At least two different paint schemes were showing through on the sides. Also notice the white "pie-shaped" sections that have been painted onto the gondolas' wheels. These triangle sections were easily seen by an engineer descending the grades on Baxter Pass, and a quick look at the car would show any wheel that had been locked up under braking conditions. This practice was not used until the mid-1930's, as another way to help with costs during the Great Depression years.

A STRANGE GROUPING of cars was photographed on the siding at Atchee during the late 1930's. This picture was taken as the photographer was looking southwest at two flatcars, followed by four-wheel caboose No. 1, a stock car, eight-wheel caboose No. 3 and the wing-type spreader—a gondola with large hinged metal plates on the side.

RODGER POLLEY COLLECTION

ROY F. BLACKBURN PHOTO – RICHARD A. RONZIO COLLECTION

CABOOSE No. 1 was one of two four-wheel "short" cabooses used by the Uintah on the line over Baxter Pass. They were extensively rebuilt, with cupolas added and more windows in the 1920's. This car was used over Baxter Pass to Watson and Dragon on most freight trains in the 1930s while caboose No. 2 was "mothballed" as traffic slowed down. She was standing next to the boardwalk in Atchee, on the siding next to the mainline, behind one of the Uintah's gondolas. The marker lamp on the end of the caboose was an oil lamp and could be moved from one end to the other by placing it in a wedge-shaped cast iron bracket that was attached to the side of the car, visible on the other end in the same location.

UINTAH CABOOSE No. 1 was used over Baxter Pass on most freight trains during the 1930's, and she was seen here on the tail end of a train of flatcars, ready to go north again. This picture was taken May 9, 1939, looking east from the siding toward the residential part of Atchee.

OTTO C. PERRY PHOTO
DENVER PUBLIC LIBRARY, WESTERN HISTORY DEPARTMENT

OTTO C. PERRY PHOTO
DENVER PUBLIC LIBRARY, WESTERN HISTORY DEPARTMENT

IN THIS PICTURE, caboose No. 3 was sitting at the far south end of the yard in Atchee on April 9, 1939. The car in the background belonged to famed railroad photographer, Otto C. Perry.

CABOOSE No. 3 was utilized mostly on the less severe section of the line between Mack and Atchee. No. 3 saw through service to Watson on occasion, but the heavy caboose usually ran behind freight trains on the flatter portion of the line, leaving the heavy grade on Baxter Pass to the Uintah's four-wheel "shorty" cabooses. Here, it was spotted at nearly the same location at different times on the main siding in Atchee.

CHRIS H. UEBERROTH COLLECTION

CHRIS H. UEBERROTH COLLECTION

APPARENTLY, THE DOOR and window frames of caboose No. 3 were painted a darker color than the body of the car. This caboose was standing in Atchee during November 1938. This car presently is being restored at the Cross Orchard Historic Site in Grand Junction, Colorado.

BINDING SPACE

STATION RECORD

Form 2097
(A. A. R. Standard Form AD-161)

______ STATION ______ 19__

CONSIGNEE ______ FREIGHT BILL NO. ______

DESTINATION ______

ROUTE ______
(POINT OF ORIGIN TO DESTINATION)

Received of The Uintah Railway Company THE FOLLOWING DESCRIBED ARTICLES IN GOOD ORDER:

WAYBILLED FROM	WAYBILL DATE AND NO.	FULL NAME OF SHIPPER	CAR INITIALS AND NO.
POINT AND DATE OF SHIPMENT	CONNECTING LINE REFERENCE	PREVIOUS WAYBILL REFERENCES	ORIGINAL CAR, INITIALS AND NO.

NUMBER OF PACKAGES, ARTICLES AND MARKS	WEIGHT	RATE	FREIGHT	ADVANCES	TOTAL
	WEIGHT SYMBOL L. C. L. FREIGHT	TOTAL PREPAID $			

LOCATION		NOTICE MAILED	DATE OF DELIVERY	TOTAL
WAREHOUSE	POST OR SECTION	19__	19__	

RECEIPT OF PROPERTY ACKNOWLEDGED

______ CONSIGNEE

PER ______

DELIVERED BY

DELIVERY CLERK

CHRIS H. UEBERROTH COLLECTION

IN THE EARLY SPRING of 1939, Nos. 21 and 20 were sitting on the so-called "dead line" in Atchee. The smokebox door had been opened for access into the boiler. The front boiler plate had been replaced during one of her many rebuilding jobs over the years. Notice the ornamental caps on the top of the sand domes and side water tanks—dating the little passenger engines to a time when steam locomotives were designed to be stylish, as well as functional. In the background, you can see gondola No. 312, with the front section of the drop-bottom undercarriage open. A large lever at the end of the gondola was geared to turn the heavy rod going along the side of the car. This rod was attached to steel mounting brackets, fitted to a chain. This, in turn, held the doors to the desired position. When the lever was rotated, the arms would lower or raise the doors, allowing the coal to drop out. Behind this gondola was one of the Uintah's tool cars, built with a cupola, a small side door (on this side) and windows at two different heights.

Form S-9--4-35--800

THE UINTAH RAILWAY COMPANY

Car No.________________ Date________________193____

BAD ORDER

Defects________________

CHRIS H. UEBERROTH COLLECTION

CHRIS H. UEBERROTH COLLECTION

OLD ENGINE No. 20 had been spotted for one last time near the end of the wye during November of 1938. Except for a missing headlight, the locomotive looked restorable.

"DEAD LINE" WAS THE NAME of the track behind the northeast side of the coal tipple during the late 1930's. This was where old engines went to die; passenger engines Nos. 20 and 21 would sit here for the last few years of the Uintah Railway's life. This picture was taken during the summer of 1938, showing a Shay coaling up at the tipple. In the middle of the picture, you can see the remains of Shay No. 2. The water tank had been removed and was being used as an auxiliary water-supply tank for the boilerhouse in Mack. The rest of No. 2 remained here, used for spare parts before being scrapped during the following year at this location. The backup light on No. 21 would never shine again, as this engine also would be scrapped, after being run down the line to Mack.

ROBERT C. BOULTER COLLECTION

ROBERT C. BOULTER COLLECTION

THE EAST SIDE of Bridge 28-A showed the positioning of the pedestrians' railing. Although the picture does not show it well, the middle of the bridge actually was closer than the ends due to the curvature of the bridge. The trestle bents for this span were square timbers, rather than the standard round uprights used for pilings on most of the Uintah's bridges. The two middle bents measured 48 feet in length and went into the mud for several feet. Notice how the stringers between the bents were attached in groups of four. The middle sections used all four for support between the bents, while the ones near the ends utilized the outside stringers as braces on the dirt. The white mound piled at the left was the location of the ash dump for the engines and boilerhouse.

ALTHOUGH TWO BRIDGES were longer, Bridge 28-A was considered to be the "biggest" trestle on the railway. To railroaders and townsfolk along the line, its size was measured not in length, but in the depth of the wash it crossed. West Salt Wash Creek. Water from many springs in the canyons above Atchee created a regular trickle of water throughout the year. However, during the springtime, there was a large increase in the flow of water, as snow from the surrounding mountains melted. Loose soil on the valley floor had a tendency to erode and slough off easily, causing bridge footings to continually need shoring up. On both sides of the trestle, rail was used for pilings, to hold back the dirt. A row of rails would be pounded into the dirt, wooden cribbing placed behind it and soil filled-in from the back, providing a much more stable slope and less erosion from rain.

FLORENCE REYNOLDS COLLECTION

ON JUNE 6, 1928, a young lady by the name of Francis Reynolds was standing on the rear platform of Business Car B-8. This Business car featured a highly polished exterior and well-furnished interior, and was only seen outside on rare occasions. Most of the time, this car was either inside the enginehouse or sitting at the end of the west stall. It was used mostly to transport management and corporate officials to the mines in Utah. On several occasions, No. B-8 was used to provide college students and academia from around the country with a tour over the impressive engineering feat of Baxter Pass. Here, the car was sitting on the wye, next to the coal chute.

ROBERT C. BOULTER COLLECTION

PHOTOGRAPHED NEAR the same location as the picture on Page 192, this view shows the east end of Atchee during the 1930's. The biggest change that had occurred was the location of the much larger boilerhouse, east of the tracks, and the track laid to the coal dump, on the back side of the building used to feed the large steam boiler. The siding built for extra storage after the sidings in the yard were filled up, or unavailable, was across Bridge 28-A.

ROBERT C. BOULTER COLLECTION

A YOUNG SCHOOL GIRL was walking along the boardwalk on the east side of Bridge 28-A. This boardwalk was wider than the smaller west side, and was normally used by foot traffic. One of the Shays was sitting at the other end of the bridge, waiting to do some switching. Since the bridge was at the north end of the wye, most engines ran out onto the trestle in order to reverse direction onto the wye. Notice how the railing on the left side of the picture used two large planks and the right side used three smaller planks.

DELL A. McCOY PHOTO

THE UINTAH RAILWAY GRADE that climbed Baxter Pass can be seen when you look northwest from Atchee in this picture taken on October 19, 1998. This scene shows the poured concrete walls of the machine shop, as it looked at the site of this repair and rebuilding facility.

THE MACHINE SHOP ruins at Atchee looked like this from the abandoned grade of the Uintah Railway, near what had been the station platform. This was photographed on October 19, 1998. The machine shop's distinctive shape was created from the concrete-pouring process when the walls were poured in layers, about the thickness of a railway tie. Instead of a highly finished surface, the walls were course-looking and had simple square corners. Large doors on the back side of the machine shop opened for a short track that ran into the middle of the shop floor. Most repairs were facilitated in the enginehouse; however, when heavy and bulky pieces of machinery had to be repaired, it was easier to roll the entire locomotive into the shop for the necessary work.

DELL A. McCOY PHOTO

LOCOMOTIVE SMOKE was rising high in the sky, as a Uintah train was climbing to the summit of Baxter Pass. During the Uintah Railway's busy years, this was an every-day occurance. The photographer was looking across Bridge 23-A from this vantage point. At this point, the train was only 1.5 miles away from Atchee "as the crow flys," but was nearly five miles away by rail!

ROBERT C. BOULTER COLLECTION

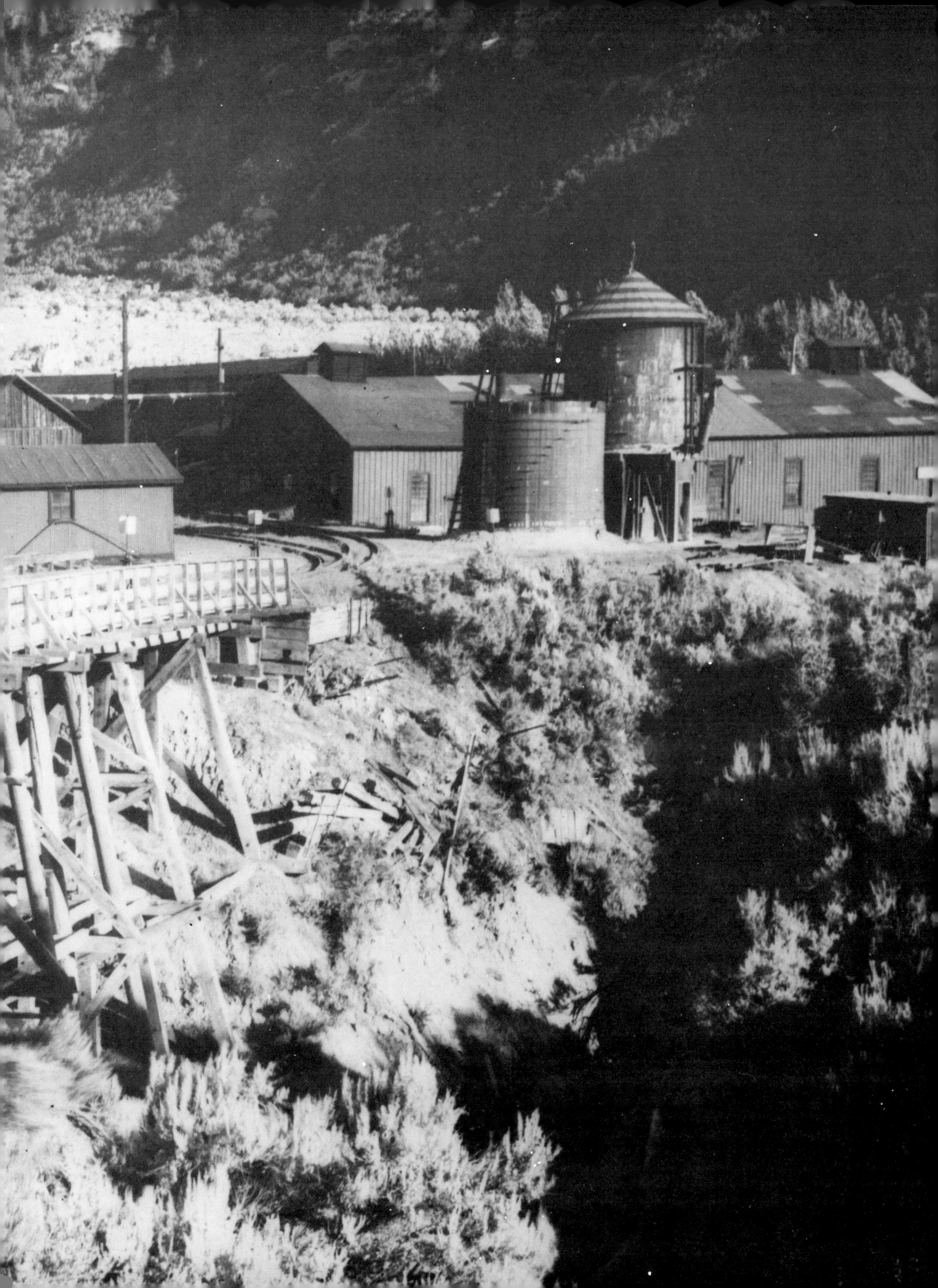

TED WURM PHOTO – CHRIS H. UEBERROTH COLLECTION

OVERLEAF: This scene at Atchee was photographed on July 16, 1939, from the northwest side of the wash, as the photographer looked back at the settlement. The patched roof of the enginehouse and stains on the water tank show the aging of the buildings in this little town. At Bridge 28-A, you can see how erosion from below the trestle caused one of the "dykes" to fall away on the far bank of the creek—down the side of the embankment into the wash.

MIDWINTER ON BAXTER PASS presented the viewer with a look at the topography that makes up the south side of the mountain side. In the foreground, camouflaged by sagebrush, was the cut in the valley made by West Salt Wash Creek. It was here, just past Bridge 28-A—to the left, off the edge of the view—that the Uintah's track made its departure from the West Salt Wash's valley, traversing a canyon en route to the pass. The string of cars and the American ditcher had been stored on the siding on the other side of the wash. This siding was added when the new boilerhouse was built, creating a need for a storage track at this location, after some trackage had been removed to make way for the new structure. To the left of these cars, you can discern one side of the cattle guard that was on the track at this point, to keep livestock from the Rogers Ranch or the high country from wandering onto the right-of-way of the Uintah Railway.

ROBERT C. BOULTER COLLECTION

TED WURM PHOTO
CHRIS H. UEBERROTH COLLECTION

AN ENLARGEMENT of the previous picture shows the boilerhouse, built between 1915 and 1919 to house a stationary steam boiler, which was used to power Atchee. The steam not only ran the machinery at the machine shop and enginehouse, but it also heated most of the town and ran the dynamo in the shop that kept the electricity flowing. The DC current from the dynamo provided the residents in Atchee with a luxury that most isolated towns of equal size only dreamed about during this same time period. In the background was a gondola, spotted on the top of the main coal-tipple track. At the left were several pieces of equipment, waiting for the scrappers to make their way into town. Both small tank engines (Nos. 20 and 21), with a gondola, tool car, caboose and the remains of the boiler of Shay No. 2 were sitting on a track called the "dead line," forgotten by everyone except the scrapping crews.

ROBERT C. BOULTER COLLECTION

DURING JULY OF 1926, brand new articulated locomotive No. 50 arrived in Mack for the first time. The original modifications performed on this 2-6-6-2T were at Mack, where the airhoses were raised to the unique Uintah position, above the couplers. After being "steamed up" at Mack, she was coupled to a load of empty flatcars and run on its first trial run up to the shops at Atchee. It was here at Atchee, at the bottom of the 7.5-percent grade, that the real test would start. General Manager Lucian Sprague and Master Mechanic Scott Shafer, who jointly designed this articulated locomotive, had their careers riding on the performance of this locomotive. Members of the Uintah's management from back East were already at Atchee, staying in the company hotel, waiting for the arrival of their "big investment." Mr. Sprague's stubborn "push" through corporate management led to the creation of the largest narrow-gauge engine produced in the U.S.A. Excitement ran high as nearly everyone in town gathered near the tracks to see the huge locomotive pull up the grade into town. Testing the new engine over Baxter Pass would start the following day, after Master Mechanic Shafer and his crew familiarized themselves with operation and maintenance procedures. Mr. Shafer (in the center), bent over the drive wheels, was looking at this remarkable piece of machinery. However, the next day's first trip up the pass was a disappointment. Encountering the extreme 66-degree curve and 7.5-percent grade at Moro Castle, the pilot truck of this narrow-gauge articulated locomotive left the rails. Mr. Sprague was riding in the cab on the trial run, and he immediately knew there were going to be stressful times ahead. After some examination of the front truck and a couple of other attempts that failed, she was backed down the grade to Atchee, unable to do the job for which she was designed! Tempers flared at the hotel in Atchee, as the "big wigs" from the East realized how much of an experiment they were a part of. Initially, Baldwin had not even wanted to do the job, thinking it was impossible to create such a large engine, and they had only relented after persuasion from Mr. Sprague and a special trip by an engineer from Baldwin to inspect the grade. With their jobs literally "on the line," Mr. Sprague and Mr. Shafer spent the next two days ironing out a plan to redesign a portion of the piping from the boiler to the front drivers, which forced the engine to bind on the sharp curve at Moro Castle, causing the derailments. With their modifications complete and sure that they would work, a celebration dinner was planned at the hotel. Ruby Luton was instructed to serve the best wine and plenty of it, for the next morning No. 50 would climb the hill—for sure. The celebration lasted late into the night, and most of the partners slept late the next morning. The management got together after breakfast to see how things were developing regarding getting the engine ready to head up the hill again. They walked outside to find Mr. Sprague and noticed that the 2-6-6-2T was turned in the wrong direction—headed away from Baxter Pass. One of them inquired as to what was the problem. "There is no problem," Mr. Sprague replied. "It's already been to the top and back!" Leaving early in the morning, Scott Shafer, Lucian Sprague and the regular train crew had taken No. 50 out of the yard and up Baxter Pass to the top, turned her on the wye and returned—*all before breakfast!*

Chapter 4

Locomotive No. 50 in Service...
ATCHEE–DIVISION POINT

ENGINEER VICTOR EARP probably was up around 5:30 a.m. on the morning of April 9, 1928, not because of an alarm clock, but more out of habit. Early during April in Atchee at 6:00 a.m. was always crisp and cold, as Mr. Earp strolled up the boardwalk toward the enginehouse office. The enginehouse and yard would have been bathed in light, as daybreak began to lighten the sky; however, Atchee remained dark, behind the large mountain to the east. Steam and smoke were pouring out of several locomotives being used to switch cars in the yard. Included was a Shay, with the rapid strokes of the three-cylinder engine unmistakable. Orders for the day were handed out from the shop office to Engineer Earp and Conductor Wehr for freight train No. 19, the regular northbound train bound for the Gilsonite mines.

On the long siding next to the mainline, their train waited for them, ready to go. In front of them was the largest narrow-gauge locomotive that had ever been built in the United States up to that time, Engine 50. Mr. Earp was the most senior engineer on the Uintah Railway and had operated this engine almost exclusively since her arrival in 1926. Mr. Lehman was the regular fireman and probably was already up in the cab. Tom Bird, one of the usual brakemen also was on hand that morning, helping to inspect the wheels and brake rigging, and the all-important air lines. The Uintah was small enough that engineers could have their own locomotive and fireman, and conductors could have a specific train; however brakemen were assigned as needed.

Articulated locomotive No. 50 was kept fired up through the weekend with the help of hostlers. On Saturday night and into the early morning hours of Sunday, several men kept the Uintah's engines coaled, sand domes filled and water tanks brim-full. Hostlers were somewhat separated from the rest of the railway men. While the normal crews used the engines to work the grades all through daytime, the hostlers spent year after year taking care of the engines all through the night-time, when everyone else in Atchee was asleep. The hostlers also were responsible for keeping the fire going under the large stationary boiler that supplied steam for the work shop, and in turn, provided power for producing electricity for the town. An unfortunate employee was fired during the 1920's for allowing the main boiler to go cold – causing an entire morning of lost production in the shop. However, on this morning No. 50 was fired up and ready to go.

Some work was going to be required this morning, as several sections of new rail had to be left at Moro Castle for the section crew on the way up the hill. The section gang would have to deal with the rail later on that afternoon. The new rail sections had been loaded on a flatcar that had a full load of crossties destined for track near Rainbow. This flatcar, plus three more loaded with lumber, an empty flatcar and the customary caboose made up the relatively light consist that morning. At 6:45 a.m., Conductor Wehr wrote down their time on the job and again at 7:00 a.m., when the train departed from Atchee on schedule.

The continuous small slide at Moro Castle made this tight curve a constant source of maintenance. The week before, locomotive No. 50 derailed at Moro Castle, and several other locomotives had similar problems lately. New rail was needed to replace the warped sections on this sharp curve. On the 7.5-percent grade at Moro Castle, Engine 50 came to a stop, with the flatcar with new rail, adjacent to a small stack of rail on the ground next to the track. A total of 30 minutes was required to unload the rail and get the train moving again. (Actually, 15 minutes were taken to unload the new rail and load two sections of worn rail of no use here.)

A routine stop was made at the top of Baxter Pass to check the airbrakes and inspect the brake rigging, standard procedure in both directions. Some 35 minutes later and 1,565 feet lower in elevation, No. 50 arrived at the wye at Wendella. Here, two more empty flatcars and three full water cars were picked up, along with an additional one a mile farther downgrade, at Sewall. A "cash fare" passenger was picked up and got a ride from Sewall to Dragon in the caboose. At Dragon City, two boxcars full of "merchandise" were added to the consist, and a water car was dropped off. Again, at Ute siding, two more empty flatcars and an empty stock car were cut into the northbound train – now totaling 16 cars, including the caboose. (Technically, the train was going "westbound" according to the Uintah Railway's timetable terminology.)

At Rainbow Junction, one of the water cars, the empty stock car and the two boxcars were cut out of the train and taken down to Watson. Then, En-

ELEV 6340	ATCHEE	MACK 28.3 M

RODGER POLLEY COLLECTION

DURING LATE 1926, this was the scene at Atchee, as the newly arrived articulated locomotive No. 50 was taking on water. The side tanks for water on the new engine created a problem for the train crews. The engine was much taller than the small passenger tank engines, and the stand pipe was inadequate with its three-inch pipe. Also, the large tank spout was too long in reach to connect on the articulated's side tank. The short-term solution was a piece of tin pipe that was run from the elbow of the water tank spout directly to the locomotive side-tank intake. With the water tank spout in the up position the tank valve was opened and the water was diverted into the added piece of pipe. This was a clumsy process that required dexterity on the part of the fireman who had to steady the pipe and open the valve at the same time. The loss of precious water was the result, as it flowed over the side of the engine. This problem later was remedied with the modification of Nos. 50 and 51, with a filler spout placed in the middle of the boiler, which allowed it to simultaneously fill both sides of the engine. The water tank glistened in the winter afternoon sun as Engine 50 was preparing to take a load of empty flatcars over Baxter Pass to the Gilsonite mines. The ladder up the backside of the tank allowed access to the interior of the tank for maintenance. Also notice the small ladder on the secondary tank, in the shadows.

ANOTHER MODIFICATION was needed early in No. 50's life. Air circulation through the cab was only allowed through two small portholes on the back of the cab, one on each side of the coal bunker. This quickly turned into a sore point with the engine crews that first summer, as the cab temperatures climbed in the hot summer sun. The heat from the boiler and firebox were bad enough; however, coupled with the summer sun beating down on the dark-colored cab, temperatures became unbearable for the engineer and especially for the fireman. Grumbling from the crew subsided with the approach of winter, and by the next spring, the small coal bunker on the back of the engine was proving to be too small for efficient operation between certain points on the railway. Also, coaling was a problem at coaling docks because there was no easy way for the crew to get to the top of the cab. The solution that next spring was to rebuild the coal bunker. The top sides of the bunker were changed from their original slanted sides to an upright side to the top of the cab. Six sections of pipe were extended from the back wall of the cab to the exterior of the bunker. These pipes dramatically helped with the air flow through the cab, and still allowed the coal to move down to the fireman's opening. Grab irons were added on top of the cab and up the left side of the rebuilt coal bunker. This allowed easier access to the top of the cab while fueling and getting water. The airpumps were moved from the pilot to locations in front of the side water tanks, next to the boiler. This eliminated the connections to the front pilot and created one less maintenance headache. The sandbox was increased in size on the front pilot allowing for not only more sand, but also providing more weight for the front end, which still had a tendency not to stay on the track, as the drivers and back truck did. Also, the paint scheme had been changed to the larger numbers and small UINTAH RY. lettering under the cab, which took place not long after delivery.

RICHARD A. RONZIO COLLECTION

ROBERT C. BOULTER COLLECTION

TRAINS WERE MADE UP using the mainline and the long siding that paralleled it, in front of the company cottages. Here, 2-6-6-2T No. 50 was switching on the mainline, while one of the short cabooses was on the siding up ahead. The strongly alkaline soil, combined with surface moisture, caused wood placed in contact with the ground to rot over the years, as witnessed by the additional supports that had been added to the bases of the electric poles next to the boardwalk.

KELLY BARLOW COLLECTION

ALL THE SHOP WORKERS wanted to get in this photograph. What had been the largest narrow-gauge locomotive ever built had just gone through modifications at the machine shop in Atchee, and the shop crew was more than proud of its achievement. The men in this view of either 1927 or 1928 were, from left to right: Charles Lyons, Bud Rogers, Elmer Collins, Gilbert Miller, Morris High, Glen Griffin, Ben Wadell, Karl Marshall, Robert A. Hester, Carroll Barlow, Mr. Earp, Paul Wagaman, Bob Rogers, Mr. Mort, John Simpson, Mr. Zellers, Mr. Kendall, Mr. Gold, Scott Shafer and Mr. Dunlap.

gine 50 returned to haul the remaining 12 cars up the steady 5.0-percent grade on the Rainbow Branch. Train 19 West pulled into Rainbow at 12:20 p.m. Switching out the water car to the end of the wye in Rainbow was the first priority, allowing its load of water to discharge into the cistern so the empty car could be taken back to Wendella for filling. Switching the empty cars out of the train and replacing them with loaded ones required an hour and 15 minutes – allowing the rear-end crew to enjoy lunch time before the engine crew could take a lunch break. So... Mallet No. 50 pulled out of Rainbow at 1:35 p.m., with seven flatcars loaded with bags of Gilsonite ore, an empty water car and the caboose.

The water car previously left at Watson now was empty and ready to return to Wendella. It was picked up, along with the other empty water car at Dragon City and four empty gondolas. The passenger from that morning was ready to return to Sewell, along with two more fares for Atchee. At Sewell the train stopped to let off a paying customer, and this 15-car train then made it to Wendella. Here, the four gondolas, three empty water cars and three loaded flatcars of Gilsonite ore were set off and two empty flatcars were added to the consist. No. 50 then set out to do the task that it was designed to do – pull loaded cars of Gilsonite ore up the 5.0-percent grade on the north side of Baxter Pass. At Columbine, there was a minor setback, as some part of the train derailed, but within 10 minutes, it was back on the track and running for the summit. Again, at the top of Baxter Pass, at an elevation of 8,437 feet, all the brakes were checked and rechecked for the 7.5-percent downgrade, with about 200,000 pounds of ore behind No. 50, each trip down the south face of Baxter Pass focused everyone's attention on safety.

Passing over the newly laid rail at Morro Castle, Engine 50 pulled into Atchee at 6:55 p.m. In another 10 minutes, they were "tied off," had handed over the train to the hostler and switching crew, and clocked out at 7:05 p.m.

On the next morning, the same crew gathered together at 6:45 a.m., and departed from Atchee at

RICHARD A. KINDIG COLLECTION

THE IMMENSE SIZE of one of the largest narrow-gauge locomotives ever produced is evident in this photograph, showing Superintendent Earp (left) and Master Mechanic Shafer standing near the pilot of No. 51. This motive power just fit inside the doors of the enginehouse and made for tight working conditions in some areas of the building. All of the work done on the two articulateds over the years was completed in this enginehouse, since the large engines could not even fit into the small area of the machine shop. The wooden cross member on top of the roof was placed there to string telegraph wires between the enginehouse, boiler plant and yard office.

the scheduled 7:00 a.m., with seven empty flatcars and the caboose. They topped "the hill," as the railroaders called it in no time, leaving the train of empties to be taken down later, and exchanging them for three flatcars loaded with ore and four empty gondolas that had been left there the night before. Turning on the wye at Wendella, No. 50 headed back up over the pass. They arrived back in Atchee at 11:35 a.m. and were clocked out at 11:45 a.m., making for a short work day and a little free time.

At 7:00 a.m. on the next morning, the crew and Engine 50 were at it again, as they pulled a consist of five cars, including the caboose, out of Atchee on April 11, 1928. The train was made up of two boxcars loaded with freight (destined for the end of the line at Watson), a flatcar load of lumber for the mines at Rainbow and two empty flatcars. At the summit of Baxter Pass, another empty flatcar, three stock cars and a flatcar piled high with cinders were added to the train. (Cinders from various locations along the line often were shoveled onto a flatcar, and boards were placed down the sides of the car, back of the wooden stakes, forming a shallow gondola. At the desired point, the boards were removed, and the cinders were allowed to "sift" off the sides of the car, as it went downgrade at the "whopping" speed of 14 miles per hour. Sometimes, these loads were accompanied by a section gang with shovels. At the place where cinders were badly needed for ballast, the section foreman had the train stop, where the crew quickly cleared the flatcar of cinders. Then, they stayed behind to work on that section of track, as the train went on downgrade to complete its run. The section between Col-

THIRTEEN YEARS SEPARATE this picture of No. 50, taken during November of 1939, and the photograph of her arrival in Mack on Page 87. Many modifications had taken place, most notably (from this angle) the relocation of the air compressors to the pilot deck, next to the smokestack, above the pilot wheels. Also observe what looks like an addition to the front of the water tank. This actually was a metal box filled with scrap iron, to help weigh down the front of the engine. When full, the large water tanks had a tendency to cause the rear drivers to track better than the front ones, and in an effort to distribute more weight to the front drivers, this box was added to No. 50. Another one, just like it, was on the pilot deck of No. 51.

CHRIS H. UEBERROTH COLLECTION

ROBERT C. BOULTER COLLECTION

A LADY PASSENGER was standing on the back platform of a combination car, in front of the machine shop at Atchee, on a cold winter day in the late 1920's. Snowstorms had left piles of the wet-and-icy stuff all over the Uintah's yard, where the men had shoveled the tracks clean. The pilot of engine No. 50 was packed full of snow, after either helping fight a snowstorm on Baxter Pass or pushing through a foot or so of the white powder on the way up from Mack. The location of the station platform near the place where the men were standing—on the right side of the train—was a precarious area during the winter for passengers. They had to carefully walk through the packed ice and rails to the Uintah Hotel, in the direction of the camera.

OTTO C. PERRY PHOTO – RICHARD H. KINDIG COLLECTION

THE HUGE ENGINE'S SIZE can be viewed here, as No. 51 covers the entire picture, hiding all of Atchee in the background, except for the water tank. The modified water intake on the engine can be seen here, as a square box on the top of the boiler, with three small pipes running to the tank. These pipes later were replaced with a single larger pipe, to aid in filling the large side tanks faster. Look at the cinder trap (screen) on the smokestack. This picture was taken from the north end of the wye in 1935. Compare the coal bunker on No. 50 with the one on this engine.

umbine and Lake McAndrews needed more cinders than anywhere on the pass, as natural springs and "seeps" were located all along this section, causing the roadbed to require continuous reworking. At the wye at Wendella, No. 50 picked up the customary three full water cars for the towns in Utah. Nos. 018, 019 and 020 were in constant use between the water tank at Wendella and Dragon, Watson and Rainbow – always trying to keep up a thirst for water that would never be quenched! Along with the three water cars, the Mallet added seven empty flatcars destined for the mines, making a long train of 20 cars and the caboose, for the downgrade to Dragon City. At Milepost 48, empty stock car No. 408 was set out on the isolated siding at Urado. A small two-car spur, Urado was a combination of the

LOCOMOTIVE No. 51 provided this appearance during the mid-1930's, after a number of functional changes to the exterior had been made. Due to the heavy grades on Baxter Pass, second steam domes were added to both of the articulateds, behind the sand domes. These were installed to ensure that the crown sheet in the boiler would be adequately cooled when the engine was descending the 7.5-percent grade. The water would run forward in the long boiler, opening the possibility that the crown sheet could overheat, causing a boiler explosion. Engine 50 had this modification made before No. 51 was delivered, and it is unclear why this change was not made in the builder's plans for No. 51. Whatever the reason, the crew at Atchee produced the addition, as well as several smaller ones over time. The railing on the front deck was extended closer to the pilot for brakemen, who constantly rode on the front deck while switching cars. A cinder screen had been added to the smokestack, and a new water-filler tube had been constructed on the center line of the boiler to aid in filling the large water tanks. The square box, with the three large pipes coming out of it, was installed on the center line of the engine because most of the water spouts on the line were designed to come down in the middle of the tracks. Since the articulateds had water-intake holes on the tanks themselves, it was a chore to fill their water tanks. The spout would be lowered, and a piece of galvanized pipe or a long canvas tube would have to run from the downspout to the side tanks. This arrangement allowed the articulateds to fill with water like the rest of the Uintah's fleet of engines and water cars. Also of interest was the number of extra grab irons added to the cab area of No. 51. A foothold on the side of the cab, a grab iron on the corner, next to the front window, and a ladder and grab irons on the coal bunker, at the back of the engine, allowed crew members easier access to places on the engine. On top of the cab, you can see a large-diameter hose and a small backup light that was used infrequently.

RODGER POLLEY COLLECTION

CHRIS H. UEBERROTH COLLECTION

THIS ARTICULATED LOCOMOTIVE had been relettered during 1936, in the same design with brighter letters and "U R Y" on the side of No. 51's cab.

words "Utah" and "Colorado." Oil exploration in the area was one reason for its existence, as small carlot loads of materials made their way in and out of the area for nearly a decade.

At Dragon, two full water cars were set out to help fill their cistern and water tank on the hill above, along with three empty flatcars. Two more empties were set out at Country Boy and three at Rector. A notation for boxcar No. 218 and stock car No. 410 was made at Country Boy on the original bill of lading, to indicate that the cars were loaded and needed movement to the south. The Gilsonite ore was trucked down to the railway from this isolated mine to Country Boy, where it was loaded on flatcars set out there for this purpose. Once loaded, they were marked on a placard to let the railway men know that they were ready to go. In this way, these two cars were made note of, and they were

CHRIS H. UEBERROTH COLLECTION

ENGINE No. 51 was taking on water as she sat on the mainline, and No. 40 was resting on the siding on a cold winter day. Every little bit of steam made a cloud, and smoke rose from the stove in combination car No. 50, at the back of No. 51's train. Notice the handcar in the foreground and the short track going off to the handcar shed to the right.

picked up that afternoon on the return trip. At Rainbow Junction, the two boxcars full of freight and the two empty stock cars were cut out and switched into Watson. The remainder of the train then was taken up the grade to Rainbow and arrived there at 12:00 noon.

At Rainbow, Engine 50 left the train to switch out the water car to the end of the wye and let it start discharging into the cistern. In the meantime, four flatcars that had already been sitting in Rainbow were coupled to the flatcar with the lumber load, and taken down the spur to the Thimble Rock Mine. Nearly as bad as any place on Baxter Pass, the short spur to the Thimble Rock Mine was just the terrain the Mallet was built for. The train would be backed down the steep incline south from the rock shelf that the town of Rainbow was built on. At the bottom of the hill, back southwest before starting another sweeping curve back to the east and over to the Gilsonite mine shaft, which had been sunk into the hill below. The four empty flatcars and the carload of lumber for the mine then were set out next to small loading trestle. This loading trestle had rail from the mine extending out parallel to the track, and near the same level as the track, to aid in loading the heavy bags of Gilsonite ore. Mallet 50 then returned with five cars loaded with Gilsonite ore, going up the hill to Rainbow.

ARTICULATED LOCOMOTIVE No. 51 was ready to head out for Baxter Pass. Some noteworthy details are the location of the vent pipes into the back of the coal bunker, which provided more air flow for the engineer and hard-working fireman. Also notice the use of the back-up light on top of the cab, as she was photographed during 1935.

RICHARD H. KINDIG COLLECTION

RODGER POLLEY COLLECTION

WATER SPLASHED over the sides when the water spout overflowed the tanks. An unidentified engineer and his wife were standing on the front pilot step of No. 51. The Mallets were thirsty for water on the big hill; with four cylinders pounding away, the 2-6-6-2T's used nearly twice as much water as most locomotives on the same section of the track.

28 A

W. C. LARSON COLLECTION

ARTICULATED LOCOMOTIVE No. 50 sat on the Baxter Pass end of Bridge 28-A, complete with a bridge-identification placard and a "No Trespassing" sign. This trestle was one of only a few on the line that had boardwalks and side rails for pedestrians. Originally, it was not built with these side rails, but they were added later—especially necessary after the relocation of the post office to Rogers' ranch, on this side of the wash. Nearly every family member would take the walk to the post office at one time or another. After the post office was moved back to the town side of the wash, to reside in Luton's store, the crossing of the bridge was discouraged, and the "NO TRESPASSING" sign was placed on the narrower west side, making the east side the only way to cross the bridge.

ROBERT C. BOULTER COLLECTION

BELOW LEFT: This portrait of the northern abutment of Bridge 28-A at Atchee was taken at 6:45 a.m. one morning, looking toward the Rogers' ranch. Ranch facilities consisted of these small buildings and white tents, at the right of the train. Beginning to strain against the ever-increasing grade ahead, black smoke poured from two hard-working engines. Notice the rocks under the north end of the bridge, which had been stacked up to help keep the soil from eroding from the pilings. In the foreground you can see the switch leading to the small spur for maintenance-of-way equipment.

THE UINTAH RAILWAY'S GRADE became increasingly steeper a short distance past Bridge 28-A, quickly becoming a 5.0-percent climb up the narrow valley west of Atchee. This photograph was taken as the photographer was looking back downgrade toward Atchee. This view was made during the middle of winter, as the deep cut through the snow indicated that this was no more than another normal winter on Baxter Pass. Atchee was located in the bottomland of the valley, just out of sight, behind the lower hills in this view, near the location of the nearest telegraph pole.

CLIFFORD HAYCOCK COLLECTION – COURTESY RODGER POLLEY

BILL AND SUE KNOUS COLLECTION

ARTICULATED LOCOMOTIVE No. 50 was making her first trip up Baxter Pass when the pilot wheels derailed at this sharp curve. Binding caused the derailment—making the maiden voyage of this new engine somewhat of a disappointment during August of 1926. Here, Uintah railroaders had gathered by the side of the big 2-6-6-2T to help put the wheels back on the rails. The attempt was made to pass this site several times that day, with the same effect—derailment! There was not enough flexibility in the running gear to make it around the extremely sharp curve without wheels coming off the track. As a result, the train had to be backed down to the machine shop in Atchee so some minor modifications could be made. This train consisted of the rarely photographed outfit car, No. T-12 (used for major derailments), and a combination car to accommodate company officials and employees that were passengers for the inaugural journey up the pass.

There, two more empty flatcars, including one used for cinders, were taken to the China Wall Mine, west of Rainbow. The now-empty water car was added to a train of 10 flatcars loaded with bags of Gilsonite ore and one empty boxcar. At 2:10 p.m., the train headed back downgrade toward Rainbow Junction.

There, it cut out the empty boxcar (No. 219) and took it to Watson. The sheep-shearing pens at Watson were a great source of traffic for the Uintah during certain times of the year, and huge bags of wool took a lot of space on each car. Although the weight was not nearly as heavy as a load of Gilsonite, the volume of the huge bags created a need for nearly as many cars. It was less expensive for the sheep men to pack the wool as tightly as possible to conserve space, so the shearers would use

(Continued on Page 315)

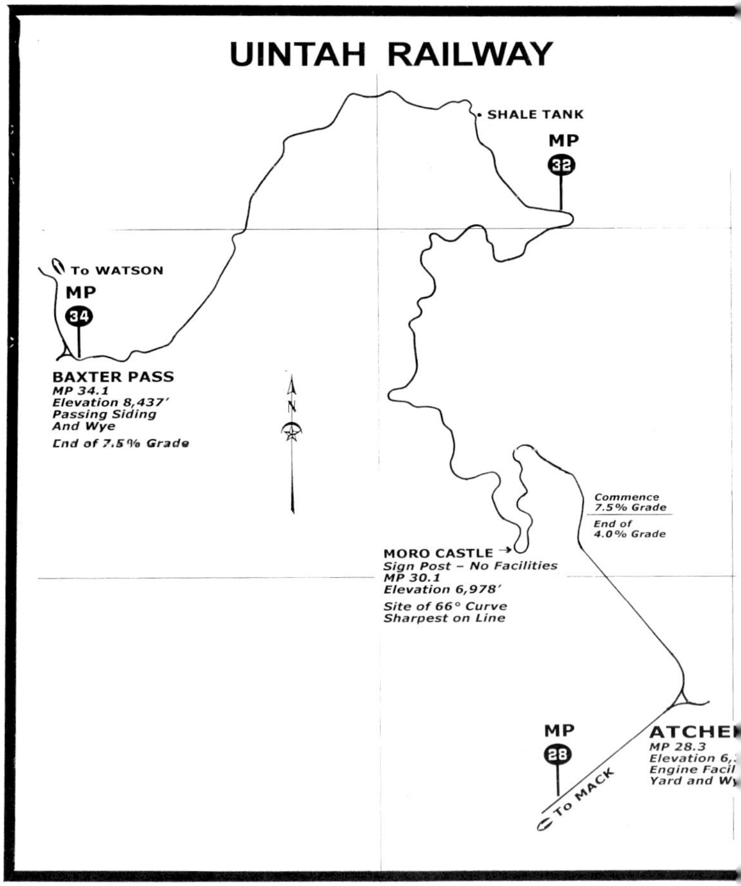

ONE OF THE FIRST ASCENTS of Baxter Pass found the new No. 50 pausing on the mainline at the west end of the summit wye. This was during the trial runs of 1925, and the engine had the Uintah's tool car in tow. This car had the dimensions of a boxcar, but was equipped with a cupola and had a caboose platform at one end.

BILL AND SUE KNOUS COLLECTION

BILL AND SUE KNOUS COLLECTION

THE AMAZING FEAT accomplished by Baldwin Locomotive Works can be seen as the pilot of engine No. 50 made its way around the famed curve at Moro Castle. The 7.5-percent grade and 66-degree curve made for a nearly impossible task for most rod engines. Here, you can see how far off the center line the pilot and front drive wheels were as the articulated engine came around the curve. The front part of the boiler was hanging out over the edge of the grade, as a brakeman walked in front of the massive, but compact, 2-6-6-2T bearing down on him. Mathematically, this curve had an 87-foot diameter, and if a model railroader fabricated this in 1:64 scale ("S" narrow gauge), it would occupy approximately 24.5 inches of table top.

OTTO C. PERRY PHOTO – RICHARD H. KINDIG COLLECTION

FROM THE AREA INSIDE THE LOOP at Moro Castle, Otto C. Perry photographed the Uintah's No. 50 climbing the 7.5-percent grade on Baxter Pass on May 9, 1939. Hauling a water car, three boxcars and combine No. 50 up the winding trackage near Morro Castle was no easy task. However, this articulated

locomotive operated considerably faster and more economically than operating the Shay geared engines, which ran on this route for 15 years before No. 50 arrived. This view shows the train as it was just below Moro Castle, on the extremely winding and steep grade that made Baxter Pass famous—*or infamous!*

OTTO C. PERRY PHOTO – RICHARD H. KINDIG COLLECTION

SEVERAL HUNDRED YARDS past the first picture in this group, Engine 50 straightened out momentarily before entering the Moro Castle curve, as the articulated tank engine rumbled past the photographer. Even as large as these locomotives were, they were dwarfed by the hillsides all around them, making the engine seem compact as she was climbing up the line. Notice in this view

the size of the two larger steam domes, in front and behind the smaller sand dome. The whistle and electric generator had been positioned behind these domes.

Continued from Page 310. . .
someone to jump up and down on the wool sacks as they were loaded, to compress the wool as tightly as possible before shipping. The long burlap sacks were placed through a large hole in a board platform, with the open end up. Then, the wool was loaded into the sacks through the open end, and someone would stand in the sack and use their feet to compress the wool as it was placed in the bag. When full, the bag was sewn shut and released from the platform and rolled down a short grade to the railway cars below.

Here, on the end of the siding at Watson, the crew dropped off the empty boxcar and picked up four boxcars and two stock cars loaded with wool. Then, they took these cars back up to Rainbow Junction and switched them into the train. No. 50 got another good workout, as she pulled a consist of 18 cars upgrade toward Dragon. At Country Boy, another boxcar and a stock car full of ore were cut into the train for the last mile upgrade to Dragon with 22 cars. At Dragon, six flatcars loaded with bags of ore were cut out, and the now-empty water cars from the morning drop off were added to the train for the return trip to Wendella.

It was here that the crew probably got the news that they would be used as an "extra" for water and would not make it back to Atchee that night. The day was already running long, but more water was needed at Dragon. Because of the way the traffic was coming out of the area that day, there was not a way to efficiently obtain water the next day, so while the motive power was on this side of the hill, the decision was made to run an extra train for water back to Dragon from Wendella that night. The train proceeded up the valley toward Wendella, picking up the stock car that had been spotted at Urado that morning, and now labled "oil." This must have been barrels of oil placed inside the stock car.

At Wendella, the three empty water cars were switched out of the train and filled from the water tank at that location. It is unclear as to how often the Mallet had to take on water, but this would be as good a place as any. From here, the "water run" was made back downgrade to Dragon, arriving there at 7:05 p.m.

Members of the train crew might have stayed at friends homes in Dragon, or they could have obtained rooms in a boarding house. However, by 1928, crews usually bunked in the freight warehouse, where there was an office and a coal stove for heating. The engine was kept hot overnight at Dragon, which was the main railway yard on the north side of the pass. Although most freight was shipped to Watson or Rainbow, Dragon had the largest capacity for cars, coal or water, and any minor service work or adjustments could be made here to the drive gear and brake system.

The crew would be running an extra train tomorrow morning, and the Mallet started switching to assemble a new train at 3:45 a.m. The three water cars had all been emptied into the cistern at Dragon during the night and were ready for another run up to Wendella. At 4:55 a.m. No. 50 departed from the yard at Dragon, hauling the extra train, consisting of the three empty water cars, eight flatcars loaded with bags of ore and the caboose.

OTTO C. PERRY PHOTO – RICHARD H. KINDIG COLLECTION

***TURNING HIS CAMERA** toward the south, Otto C. Perry shot this final picture in the series, as the Uintah mixed train entered the 66-degree curve at famous Moro Castle...*

UINTAH BRAKEMEN had to be surefooted as they walked along the tops of the freight cars—swaying from side to side as the train was winding its way over an unstable section of track. The large shelf of sandstone that protruded from the mountain was a natural location for ground water to flow from the hillside. This effect was aggravated with the Uintah's grade that cut into the mountain at Moro Castle. This created a constantly flowing "seep" that caused the area to constantly push down the hillside. This problem

was so bad by the 1920's that a short siding was laid here, and a tool shed was built for the section men required to rework the grade at this curve on a regular basis. On this warm spring day, the weather must have been nice, as the back door of combination car No. 50 had been left open, showing the afternoon sunlight shinning through the windows onto the floor down the car's center isle.

Arriving at Wendella, the crew dropped off the train and coupled up to the consist they had left the night before. Two stock cars and four boxcars loaded with wool fleece from the shearing station in Watson, a stock car full of mysterious "oil" from Urado, four flatcars, three boxcars and a stock car loaded with bagged ore made up this train. A total of 15 cars, plus the caboose, made this a heavy train for Baxter Pass. Nevertheless, this was not a problem for No. 50, as the crew was back in Atchee at 11:40 a.m. They let the brakes cool on the articulated, welcomed the comforts of home and got some much-needed sleep – making up for the long work day and short night at Dragon, Utah.

On Friday, April 13, 1928, the same crew put together Train 19 West just like every other day. Three boxcars destined for Watson were loaded with freight for Vernal and two empty flatcars made up the consist. Conductor Wehr had to record the seal numbers from the metal bands on the sealed boxcars after they were loaded at Mack. These numbered bands were meant to prevent theft, placed on doors and hatches. These metal bands were numbered for identification and listed on the paper work.

This train went up to the summit of Baxter Pass, where six more empty flatcars were added to the consist, before the Mallet headed on down the north side of the pass. These flatcars probably had been left there the day before by the "hill extra" that ran from Atchee to the top of the pass. These extra runs were powered by Shay geared engines. The Shays were great for hauling loaded cars up Baxter Pass, but were slow for the runs on either side of the pass – which made them inefficient when compared with the Mallet locomotives. This very month the fate of the Shays would be sealed, as the Uintah received Mallet No. 51 fresh from the Baldwin erecting plant. These two 2-6-6-2T locomotives would spell the end of the reign of the Shay engines on Baxter Pass, and by the end of the year, three of the six Shays on the roster would be sold or scrapped.

With the empty flatcars added to the consist and the routine brake inspections complete, No. 50 started down the grade toward Lake McAndrews and Wendella. At Wendella, the normal three filled water cars were waiting, along with an empty boxcar. This train then was taken down to Watson and Rainbow – switched in the normal way – and the return trip consisted of Gilsonite ore and wool fleece. It was a long day, as the crew did not reach Atchee until 7:15 p.m. that night, where they handed over the train to a hostler and switch-engine crew.

Saturdays had a somewhat different schedule for the freight crew, as the train ran as an extra. Leaving at 5:25 a.m., the crew knew that they would be off for the weekend as soon as this extra train completed its task. The Mallet probably pounded up the hill with urgency, as three boxcars loaded with freight and two empty flatcars were taken over the hill to Wendella, to be exchanged for six flatcars loaded with sacks of ore. No. 50 returned to Atchee by 9:40 a.m. that morning! After 4.5 hours, the week finally was over, and the crew could enjoy the next day and a half off work before starting the entire process over again on Monday morning.

~ INDEX ~

**** Indicates photograph and/or caption***

— INDEX Continued —

RODGER POLLEY COLLECTION

RODGER POLLEY was born and raised in Rangely, Colorado, not far from the Uintah Railway's terminus at Dragon, Utah. Rodger went to Mesa State College in Grand Junction, Colorado, where he earned a BA degree in Business Administration. At this time, he makes Grand Junction his home with his wife Debbie. He is a past president of the Rio Grande Chapter of the National Railway Historical Society. Currently, Rodger manages a retail motorcycle store.

***THIS IS MY FIRST ATTEMPT** at any kind of writing, and at the same time, this has been my passion for several years. Please try not to be too harsh — and I hope this work lives up to everyone's expectations.*

THE WHOLE BASIS for this project has been to shed light on the history of the Uintah Railway, especially through never-before-published photographs. The picture captions included with the photographs make up the greater part of my project, provided for you to peruse as you view the illustrations from long, long ago. I have not tried to capture all the detailed history that was encompassed in Henry Bender's well laid-out chronological history, *"UINTAH RAILWAY: The Gilsonite Route."* Nevertheless, it is my hope that through the many previously unseen photographs, maps and added little stories, this pictorial work will be good at "filling-in the holes" that history left behind.

When I was a teenager, my original interest came from model railroading. Throughout working on the ***UINTAH RAILWAY PICTORIAL*** project, I have kept an eye out for details that might seem fascinating for the modeler. At the same time, I have always been intrigued with the life and times the people of my parent's generation lived, especially Great Depression Era Americana. These two "target areas"— along with my intrepid ghost-town explorer fixation — have always been my principal interests. It is my hope that the mainstream railroad historians will also enjoy this publication — after all, the Uintah Railway was one of the "coolest" narrow-gauge railroads ever built.

If you have photographs, printed material or information pertaining to the Uintah Railway, I would appreciate correspondence regarding the subject. My address is: 411 Teller Avenue, Grand Junction CO 81501.

—— Rodger Polley